Global Manufacturing Technology Transfer

Africa–USA Strategies, Adaptations, and Management

Industrial Innovation Series

Series Editor

Adedeji B. Badiru

Air Force Institute of Technology (AFIT) – Dayton, Ohio

PUBLISHED TITLES

Carbon Footprint Analysis: Concepts, Methods, Implementation, and Case Studies, *Matthew John Franchetti & Defne Apul*

Cellular Manufacturing: Mitigating Risk and Uncertainty, *John X. Wang*

Communication for Continuous Improvement Projects, *Tina Agustiady*

Computational Economic Analysis for Engineering and Industry, *Adedeji B. Badiru & Olufemi A. Omitaomu*

Conveyors: Applications, Selection, and Integration, *Patrick M. McGuire*

Culture and Trust in Technology-Driven Organizations, *Frances Alston*

Global Engineering: Design, Decision Making, and Communication, *Carlos Acosta, V. Jorge Leon, Charles Conrad, & Cesar O. Malave*

Global Manufacturing Technology Transfer: Africa–USA Strategies, Adaptations, and Management, *Adedeji B. Badiru*

Guide to Environment Safety and Health Management: Developing, Implementing, and Maintaining a Continuous Improvement Program, *Frances Alston & Emily J. Millikin*

Handbook of Emergency Response: A Human Factors and Systems Engineering Approach, *Adedeji B. Badiru & LeeAnn Racz*

Handbook of Industrial Engineering Equations, Formulas, and Calculations, *Adedeji B. Badiru & Olufemi A. Omitaomu*

Handbook of Industrial and Systems Engineering, Second Edition, *Adedeji B. Badiru*

Handbook of Military Industrial Engineering, *Adedeji B. Badiru & Marlin U. Thomas*

Industrial Control Systems: Mathematical and Statistical Models and Techniques, *Adedeji B. Badiru, Oye Ibidapo-Obe, & Babatunde J. Ayeni*

Industrial Project Management: Concepts, Tools, and Techniques, *Adedeji B. Badiru, Abidemi Badiru, & Adetokunboh Badiru*

Inventory Management: Non-Classical Views, *Mohamad Y. Jaber*

Kansei Engineering — 2-volume set

- Innovations of Kansei Engineering, *Mitsuo Nagamachi & Anitawati Mohd Lokman*
- Kansei/Affective Engineering, *Mitsuo Nagamachi*

Kansei Innovation: Practical Design Applications for Product and Service Development, *Mitsuo Nagamachi & Anitawati Mohd Lokman*

Knowledge Discovery from Sensor Data, *Auroop R. Ganguly, João Gama, Olufemi A. Omitaomu, Mohamed Medhat Gaber, & Ranga Raju Vatsavai*

Learning Curves: Theory, Models, and Applications, *Mohamad Y. Jaber*

Managing Projects as Investments: Earned Value to Business Value, *Stephen A. Devaux*

Modern Construction: Lean Project Delivery and Integrated Practices, *Lincoln Harding Forbes & Syed M. Ahmed*

Moving from Project Management to Project Leadership: A Practical Guide to Leading Groups, *R. Camper Bull*

Project Management: Systems, Principles, and Applications, *Adedeji B. Badiru*

PUBLISHED TITLES

Project Management for the Oil and Gas Industry: A World System Approach, *Adedeji B. Badiru & Samuel O. Osisanya*

Quality Management in Construction Projects, *Abdul Razzak Rumane*

Quality Tools for Managing Construction Projects, *Abdul Razzak Rumane*

Social Responsibility: Failure Mode Effects and Analysis, *Holly Alison Duckworth & Rosemond Ann Moore*

Statistical Techniques for Project Control, *Adedeji B. Badiru & Tina Agustiady*

STEP Project Management: Guide for Science, Technology, and Engineering Projects, *Adedeji B. Badiru*

Sustainability: Utilizing Lean Six Sigma Techniques, *Tina Agustiady & Adedeji B. Badiru*

Systems Thinking: Coping with 21st Century Problems, *John Turner Boardman & Brian J. Sauser*

Techonomics: The Theory of Industrial Evolution, *H. Lee Martin*

Total Productive Maintenance: Strategies and Implementation Guide, *Tina Agustiady & Elizabeth A. Cudney*

Total Project Control: A Practitioner's Guide to Managing Projects as Investments, Second Edition, *Stephen A. Devaux*

Triple C Model of Project Management: Communication, Cooperation, Coordination, *Adedeji B. Badiru*

FORTHCOMING TITLES

3D Printing Handbook: Product Development for the Defense Industry, *Adedeji B. Badiru & Vhance V. Valencia*

Company Success in Manufacturing Organizations: A Holistic Systems Approach, *Ana M. Ferreras & Lesia L. Crumpton-Young*

Design for Profitability: Guidelines to Cost Effectively Management the Development Process of Complex Products, *Salah Ahmed Mohamed Elmoselhy*

Essentials of Engineering Leadership and Innovation, *Pamela McCauley-Bush & Lesia L. Crumpton-Young*

Handbook of Construction Management: Scope, Schedule, and Cost Control, *Abdul Razzak Rumane*

Handbook of Measurements: Benchmarks for Systems Accuracy and Precision, *Adedeji B. Badiru & LeeAnn Racz*

Introduction to Industrial Engineering, Second Edition, *Avraham Shtub & Yuval Cohen*

Manufacturing and Enterprise: An Integrated Systems Approach, *Adedeji B. Badiru, Oye Ibidapo-Obe & Babatunde J. Ayeni*

Project Management for Research: Tools and Techniques for Science and Technology, *Adedeji B. Badiru, Vhance V. Valencia & Christina Rusnock*

Project Management Simplified: A Step-by-Step Process, *Barbara Karten*

A Six Sigma Approach to Sustainability: Continual Improvement for Social Responsibility, *Holly Allison Duckworth & Andrea Hoffmeier Zimmerman*

Global Manufacturing Technology Transfer

Africa–USA Strategies, Adaptations, and Management

Adedeji B. Badiru

CRC Press is an imprint of the
Taylor & Francis Group, an **informa** business

CRC Press
Taylor & Francis Group
6000 Broken Sound Parkway NW, Suite 300
Boca Raton, FL 33487-2742

First issued in paperback 2019

ISBN-13: 978-1-4822-3553-1 (hbk)
ISBN-13: 978-0-367-37754-0 (pbk)

Visit the Taylor & Francis Web site at
http://www.taylorandfrancis.com

and the CRC Press Web site at
http://www.crcpress.com

Dedication

To Alexa Folashade Badiru for love, heritage, and inspiration

Contents

Preface

Global interfaces are essential for national advancement and sustainability of economic vitality. Technology is the primary foundation for global interfaces that provide benefits to all parties involved. Global manufacturing technology transfer presents strategies, adaptation methods, and management techniques for connections between Africa and the United States. The premise of this book is to advance existing technical relationships through new and updated approaches.

Africa is emerging as a new hub for international manufacturing as a result of rising costs in traditional outsourcing destinations in Asia. Many manufacturers are returning previously outsourced manufacturing operations to their home countries; however, they continue to be interested in exploring new outsourcing options. Africa is therefore poised to assume a leading role in global manufacturing. However, the foundation for doing so successfully will depend on the development and sustainability of a reliable manufacturing infrastructure in Africa. A key requirement for this will be new and innovative mechanisms for bilateral technology transfer. Although this book focuses specifically on Africa–USA manufacturing technology transfer, it is equally of direct interest to all manufacturers in the world. The contents are applicable to different audiences globally. Contents include project management, product development, technology justification, and technology transfer modes.

Adedeji B. Badiru
Air Force Institute of Technology, Dayton, Ohio

Acknowledgments

A book of this nature is a multiparty and global endeavor. First and foremost, I thank my wife Iswat, my home chief of staff, who typed many of the materials and also provided administrative oversight for the preparation of the manuscript. Her support and love over many decades in all my literary pursuits are greatly appreciated. I thank my son, Omotunji (TJ), for assisting in drawing many of the original figures used in this book. I thank Annabelle Sharp, whose sharp sense of mission contributed to the typing and paperwork for the manuscript. I thank Dr. Miguel Marquez of Venezuela, a former soccer teammate, PhD graduate of the University of Oklahoma, and a professional colleague, who, as the chairman of the 2004 Conference on CAD/CAM, Robotics, and Factory of the Future (in San Cristobal, Venezuela), gave me the permission to reuse materials from the conference proceedings. I thank Professor Pius Egbelu, who not only provided executive-level encouragement and support for this project but also wrote a chapter for the manuscript. All intellectual contributions and sweat-equity support are humbly acknowledged and appreciated.

Author

Adedeji B. Badiru is the dean and senior academic officer for the Graduate School of Engineering and Management at the Air Force Institute of Technology (AFIT), Dayton, Ohio. He holds a BS degree in industrial engineering, an MS degree in mathematics, an MS degree in industrial engineering from Tennessee Technological University, and a PhD in industrial engineering from the University of Central Florida. He is responsible for planning, directing, and controlling all operations related to granting doctoral and master's degrees, professional continuing cyber education, and research and development programs. Badiru was previously professor and head of systems engineering and management at AFIT; professor and department head of industrial & information engineering at the University of Tennessee, Knoxville, Tennessee; and professor of industrial engineering and dean of University College at the University of Oklahoma, Norman, Oklahoma. He is a registered professional engineer, a certified project management professional, a fellow of the Institute of Industrial Engineers, and a fellow of the Nigerian Academy of Engineering. His areas of interest include mathematical modeling, systems efficiency analysis, and high-tech product development. He is the author of more than 30 books, 35 book chapters, 75 technical journal articles, and 115 conference proceedings and presentations. He also has published 30 magazine articles and 20 editorials and periodicals. He is a member of several professional associations and scholastic honor societies. Badiru has won several awards for his teaching, research, and professional accomplishments.

chapter one

Systems view of the manufacturing world

The entire world is now a global system of systems. We are all interconnected through one medium or another. Global manufacturing is, indeed, one of the influential systems within the world system. Given the way we are now intertwined and interconnected around the world, we must understand our interdependency. This book presents practical modern strategies that nations, cities, and communities can adopt for developing and sustaining manufacturing for the purpose of sustainable industrialization. Although the subtitle focuses on Africa–USA technology transfer strategies, adaptations, and management, the contents are broadly pertinent for all nations, developed or developing, as a path toward economic development through manufacturing.

The economic development and industrialization of any nation depend on solid foundations of manufacturing. Over the passage of time, industries come and go and requirements for economic commerce shift. Nations that survive and continue to thrive use adaptive strategies that meet current and future needs. If good sustainable strategies are available, organizations and communities can survive any turbulence that develops in the normal course of economic development. This book is operationally focused, with adaptive contents to fit the needs of different global audiences. The book uses the case and template of trade relations of Africa and the United States to demonstrate how modern technology innovation strategies can be developed and harmonized to keep a nation moving forward economically. The Africa–USA technology transfer models and techniques are applicable to other technology transfer relationships between any pair of collaborators. Although Africa is a continent and the United States is a country, trade and commerce between the two entities are similar to any country-to-country relationships. Indeed, within the Africa–USA relationship resides country-by-country interfaces of the United States with specific countries of Africa. Some examples include the bilateral relationships between the United States and countries such as Nigeria, Liberia, South Africa, Sudan, Ghana, and Egypt. The Sub-Saharan countries of Africa are more noticeable when the issues of developing and underdeveloped countries are covered in the literature or the popular press. The economic size of the United States does, indeed, put that country on a comparative par with the continent of Africa.

There are several global issues affecting how nations interact politically, socially, militarily, economically, and commercially. Bhargava (2006) addresses many of the global issues faced by citizens around the world.

MANUTECH foundation

The background and foundation for this book are traced to the Africa–USA International Conference on Manufacturing Technology (MANUTECH) organized between 1993 and 2004 by a group of Nigerian engineers in diaspora (Nnaji et al. 1993, 1994, 1996, 1997, 2000; Badiru et al. 2002, 2004; Okogbaa et al. 2007). The conference convened on a rotational basis between locations in Africa and the United States. MANUTECH 2007 exists only in a planning document (Okogbaa et al. 2007) as the last conference that was planned but was not convened due to wide professional reassignments of the organizers at that time. The topics addressed by the MANUTECH series are still germane today for the purpose of technology transfer strategies between the United States and Africa.

At the conference, engineers, scientists, managers, researchers, chief executives, administrators, and students from across Africa and the United States participated in bilateral technology-related discussions. This book represents a continuation of the ideals established and practiced over the past several years as a result of the MANUTECH initiative. The original MANUTECH members shown in Figure 1.1 have all

Figure 1.1 Leaders and organizers of MANUTECH 1993. From left to right: Professors Adedeji Badiru, Geoffrey Okogbaa, Bart Nnaji, Pius Egbelu.

gone on to make contributions through teaching, consulting, publishing, administration, business ventures, and pseudopolitical pursuits, just as many other Africans are doing.

The economic advancement being experienced by Africa in the current global economic climate is exactly what the MANUTECH conference organizers envisioned in January 1993. It has taken two decades, but slow and steady wins the race. Africa's economic rise in 2014 is a testimony to solid intellectual foundations bearing fruit in the long run. The MANUTECH 2004 welcome statement as shown in Box 1.1 aptly conveys this optimism.

Today, MANUTECH continues to represent an international congregation of scholars and practitioners interested in the advancement of manufacturing technology in Africa. Forward and reverse manufacturing technology transfers between the United States and Africa are imperatives for global competitiveness in the new world marketplace for these partners. The collaborative theme epitomizes a triangular relationship between academia, industry, and government establishments. This facilitates pragmatic approaches to transforming classroom theories into practical application for the betterment of industry. Through the conference of the 1990s and the early 2000s, the knowledge-driven pursuits of academia were coupled with the profit-driven objectives of industry to provide a framework for the policy-driven strategies of government agencies. The wide variety of topics befitting the theme of the conference and the ensuing intellectual pursuits paved the way for several subsequent constructive pursuits by the participants.

The problems of developing countries, such as those in Africa, cannot be successfully addressed until those countries are able to implement appropriate technologies for infrastructure development and for basic manufacturing. In particular, the adaptations of existing and appropriate manufacturing technologies are essential to sustainable industrial growth and development.

Economic, social, and political progress in Africa can be enhanced by the creation of local industries which provide employment for the growing population, cater to the needs of the population for local consumer goods, provide export products, and stimulate foreign investment. This book provides a forum for the exchange of information and ideas between professionals in Africa and those in the Western world for manufacturing technology transfer strategies. Such exchanges could form the basis for creating better awareness of the role of manufacturing technologies in the context of the development and economic stability of Africa and could also become the catalyst for increased and sustained dialogue on those issues.

BOX 1.1 MANUTECH 2004 WELCOME STATEMENT

On behalf of the Network of African Engineers and Scientists, the US National Science Foundation, Nigerian Ministry of Science and Technology, Nigerian Ministry of Industry, Nigerian Ministry of Education, the University of Tennessee, the University of Pittsburgh, the University of South Florida, Louisiana State University, and the several private individuals and company sponsors, I welcome you to the 7th Africa–USA International Conference on Manufacturing Technology (MANUTECH 2004) in Port Harcourt, Nigeria.

Once again, Engineers, Scientists, Managers, Researchers, Chief Executives, Administrators, and Students are invited to participate in this important technological forum. This conference is co-sponsored by both public and private entities around the world. MANUTECH 2004 represents an international congregation of scholars and practitioners interested in the advancement of Manufacturing Technology in Africa. Forward and reverse manufacturing technology transfers between the USA and Africa are imperatives for global competitiveness in the new world marketplace for these partners. The conference epitomizes a triangular relationship between academia, industry, and government establishments. This facilitates pragmatic approaches to transforming classroom theories into practical applications for the advancement of industry.

At this conference, the knowledge-driven pursuits of academia are coupled with the profit-driven objectives of industry to provide a framework for the policy-driven strategies of government agencies. A wide variety of topics befitting the theme of the conference are presented in the conference program. The problems of developing countries, such as those in Africa, cannot be successfully addressed until those countries are able to implement appropriate technologies for infrastructure development and for basic manufacturing. In particular, the adaptations of existing and appropriate manufacturing technologies are essential for sustainable industrial growth and development.

Economic, social, and political progress in Africa can be enhanced by the creation of local industries which provide employment for the growing population, cater to the needs of the population for local consumer goods, provide export products, and stimulate foreign investment. This conference will provide a forum for the exchange of information and ideas between professionals in Africa and those in the western world. Such exchanges could form the basis for creating better awareness of the role of manufacturing technologies in the context of the development and economic stability of Africa and could also become the catalyst for increased and sustained dialogue on those issues. The forum offered by MANUTECH 2004 can generate development initiatives that can serve as pragmatic models for other developing parts of the world.

We hope that you will enjoy the conference and the host city of Port Harcourt. In between the conference activities, we hope you will find time to visit many places in the city and witness the cultural, economic, political, and social excellence and diversity offered by the local community. Once again, welcome to MANUTECH 2004. Enjoy the conference.

With the highest professional regard,

Professor Adedeji B. Badiru
General Chair, Africa–USA International
Conference on Manufacturing Technology
MANUTECH 2004, July 12–14, 2004.

Appropriate technology

Ibidapo-Obe (1996) addresses the question of whether we can develop models that will measure the effectiveness or appropriateness of technologies, keeping in mind all the facets and modes of local and national aspirations. He contends that we are in a period of rapid social and economic change, much of which is driven by science and technology. Around the world, waves of science- and technology-based innovation are leading to a multitude of new products, processes, and services. Science and technology, in effect, are merging as the basis for national comparative advantage. Whether a country produces technology or uses it strategically is dependent on an environment that is rich in specialized infrastructure, specialized technical and financial resources, educational programs, management practices, marketing services, and highly skilled workers. In order to be a part of the change process, every community must develop a *technology* engine to link with this rapid and evolutionary development. Further, Ibidapo-Obe (1996, p.194) asks, "Who will be the players in the community technology engine?" He provides suggestions that include the following:

- The technology sector
- Every local industry and business, including general business associations
- Educational, research, and scientific institutions
- Planners and officials from all levels of government

- Individuals involved in technology transfer, such as brokers, institutional representatives, and consulting engineers
- Financiers, bankers, and local investors
- Labor groups
- Elected officials, political representatives, and government officials

The entities identified above are responsible for providing the leadership and guidance that communities need to fire up the technology engine. We need to embrace information technology, which has become the engine of our modern society. Information technology is the pathway for achieving development objectives. We can now attempt to quantify and, thereafter, input all the above variables into the process of developing the technology input–output model. As a strategy, Ibidapo-Obe (1996) proposes a procedure to understand the workings of the various human sensors and activators (neurocontrol mechanisms), especially vision and the brain. These will serve as a micromodel for the larger world systems, provided we can establish a one-to-one relationship between the micro (neurocontrol) model and the macro (world development) system. The basic idea is to introduce some formalities in the restructuring of vagueness in many decision problems. Some of the considerations in this development process include the following:

1. *Inputs*: Knowledge, capital, materials, hardware, software, and human resources
2. *System*: Engineering, production, personnel, manufacturing, management, and facilities
3. *Output*: Better products and services, higher profits through technology applications

Engineering is the connecting arm of technology in the development between the current scenario and the desired future state. The products of the modern world only function as long as large parts of the society operate according to the laid out plan. Unfortunately, in many developing countries, things often don't work according to plan. Underutilized equipment, rusting machinery, and idle factories working at less than half of their designed capacities provide cogent testimonies of their dilemma. In all of the above, having solid global situational awareness is essential.

Global situational awareness

Everything is industrially interconnected in the modern marketplace. It is a systems world and all organizations must demonstrate a systems view of the world when managing international projects. Teamwork and mutual understanding take on a different flavor when different cultures are involved. Not only must we think globally, we must also act globally through culturally sensitive project management. It is impossible for any organization to run any large project nowadays without some aspect of international involvement.

World systems model

Figure 1.2 illustrates the author's systems view of the world using the Vee systems engineering model. It is seen that several factors revolve around the manufacturing life cycle. A decomposition of the overall manufacturing system facilitates collaborative manufacturing involving different players from different parts of the world. A good example is the globally connected network of manufacturers involved in building Boeing's 787 Dreamliner airplane. This interconnectivity means that manufacturers must work with industrial collaborators spanning different cultures, languages, religions, customs, trade practices, skills, and politics. The conglomeration of manufacturing works for Boeing because the company uses a global systems engineering approach.

Manufacturing project management

Manufacturing project management is the pursuit of organizational goals within the constraints of time, cost, and performance expectations. The principles and techniques of conventional project management are

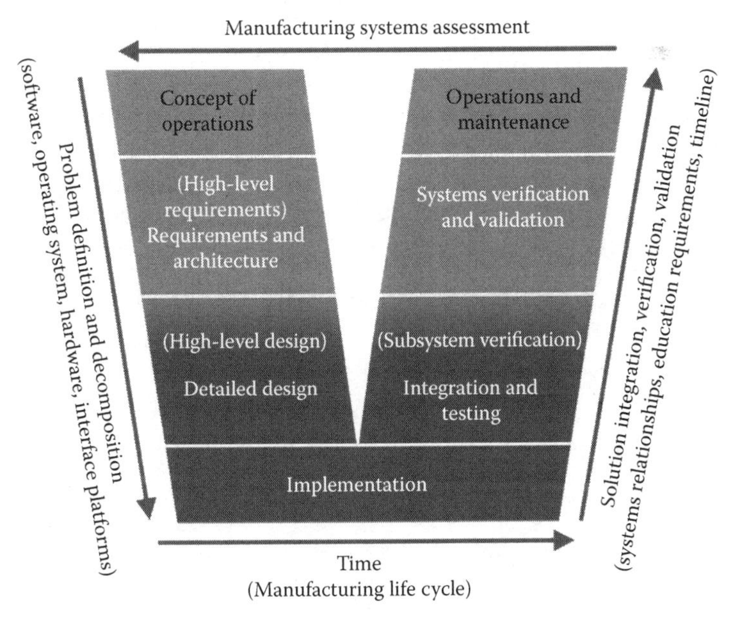

Figure 1.2 Vee systems engineering model.

applicable to the management of technology transfer efforts. Project expectations can be in terms of physical products, service, and results. Global interactions of business and industry imply that global situational awareness be instituted for international project management. Situational awareness is the process of recognizing and appreciating unique factors that define the operating characteristics in a given geographical location. These factors can range widely from one country to another. What are believed to be normal operating conditions in home country operations may be taboo, illegal, restricted, or forbidden in another country. Far too often, organizations don't pay enough attention to this fact. Training programs, briefings, seminars, and sensitivity role plays are often used to prepare personnel for overseas assignments. But the fact remains that global awareness failure points still exist among organizations operating overseas, away from home base—whether from the East to the West or from the West to the East or elsewhere.

In world news, headlines abound on topics of global project interconnections in oil, autos, industrial credits, construction, and so on. Such mega deals are not limited to the G8, G10, G12, or G20 countries. Even the small non-G economies are in the mix of international projects. What connects all of these is global project management.

Hybridization of cultures

The increased interface of cultures through international project outsourcing is gradually leading to the emergence of hybrid cultures in many developing countries. A hybrid culture derives its influences from diverse factors, where there are differences in how the local population views education, professional loyalty, social alliances, leisure pursuits, and information management. A hybrid culture poses a big challenge to managing internationally outsourced projects. But the problem can be mitigated by a strategic program of situational awareness. Table 1.1 presents some of the pros and cons of cultural enmeshing at the global level.

Table 1.1 Pros and Cons of Global Hybrid Cultures

Pros	Cons
• Global awareness for competition • Culturally diverse workforce • Access of international labor force • Intercultural peace and harmony • Facilitation of social interaction • Closing of trade gaps	• Subjugation of one culture to the other • Cultural differences in work ethics • Loss of cultural identity • Bias and suspicion • Confusion about cultural boundaries • Nonuniform business vocabulary

Market competition

Many Western industries cannot compete globally on the basis of labor cost, which is where the pursuit of competitiveness is often directed. The competitive advantage for many manufacturers will come from appropriate infusion of technology into the enterprise. Strategic research, development, and implementation of technological innovations will give manufacturers the edge needed to successfully compete globally. In spite of the many decades of lamenting about the future of manufacturing, very little has been accomplished in terms of global competitiveness. Part of the problem is the absence of a unified project management approach. Managing global and distributed production teams requires a fundamental project systems approach.

One recommendation here is to pursue more integrative linkages of technical issues of production and the operational platforms available in industry. Many concepts have been advanced on how to bridge the existing gaps. But what is missing appears to be a pragmatic project-oriented road map that will create a unified goal that adequately, mutually, and concurrently addresses the profit-oriented focus of practitioners in industry and the knowledge-oriented pursuits of researchers in academia. The problems embody both scientific and management issues. Many researchers have not spent adequate time in industry to fully appreciate the operational constraints of industry. Hence, there is often a disconnection between what research dictates and what industry practice requires. An essential need is the development of a global project road map. One aspect that is frequently ignored in this respect is the set of human-cultural factors operating in globally distributed work teams. A culturally sensitive project management approach will enable an appreciation of this crucial component of global projects.

Global awareness questions

In order to enhance global awareness, addressing specific sets of questions in each and every overseas assignment case would be more effective and sustainable, without making any prior assumptions. Even previous international experience can become out of date because local situations can change over time. The questions categorized below can be helpful.

Sample questions for pre-assignment planning

- What is the organizational mission in the overseas location?
- Are personnel aware of the global situation as it relates to the country or region of operation?
- Will the assignment be as a team, individual operation, or organizational attachment?

- What legal status will be associated with the project during the overseas assignment?
- What country locations are involved in the assignment?
- What healthcare situations and medical services are available during the assignment?
- What personal family arrangements and precautions are relevant for the overseas assignment?
- What prudent precautions are possible? Will update? Health insurance? Medical records?
- What is the currency exchange rate? Location and access to bureau de change?

Sample questions for local conditions

- What kind of local environment will be in effect as the assignment location? Urban city? Rural? Jungle? Mountain? Desert?
- What is the typical weather pattern? Seasons? Cold? Hot? When, where, and how long?
- What are the general living conditions? Access to clean water? Bathroom facilities?
- What are the domestic transportation options? Cost? Reliability? Access? Rivers? Trails? Trains? Shuttle buses? Commuter planes?
- What is the level of environment consciousness?
- What is the local social hierarchy?

Sample questions for cultural nuances

- Is there a compilation of acceptable and unacceptable behavior in the assignment location?
- What are the political beliefs?
- What are the typical customs?
- What are the prevailing religious beliefs and restrictions?
- Are there guidelines for use of body language and gestures?
- What are the offensive symbols and words to avoid?

Sample questions for geographic location

- What are the bordering regions, areas, and/or countries to the assignment location?
- What is the operating terrain? Flat? Hilly? Road system?
- Is there access to rivers, oceans, beaches, and so on?
- What is the vegetation like?
- What are the major industries? Manufacturing factories? Service industry?
- Are there farming establishments? Access to fresh foods?

Sample general questions

- What is the social environment?
- Is there nightlife? Is it accessible? Is it safe?
- What is the crime statistics?
- What is the mode of law enforcement?

- What is the restriction on currency importation and exportation?
- What is the local manpower level? Skills? Availability? Cost? Dependability?
- What is the educational opportunity? Access? Affordability? Level?
- What international support organizations are available? Red Cross?
- What is the human rights record of the locality?

If questions such as the above are addressed forthrightly, international projects can run more effectively, more efficiently, and more successfully. Typical sources of barriers to comprehensive global awareness include the following:

- *Denial*: This presents the danger of not fully recognizing the problem of global awareness. An organization may dismiss the problem or minimize the gravity of the problem.
- *Past experience*: This has the danger of luring the personnel into an erroneous decision based on a past or similar experience.
- *Complacency*: This has the danger of a feeling of overconfidence about knowing what is obtained in the assignment location, thereby dismissing the necessity for proactive preparation.
- *Insufficient information*: This has the danger of misinterpretation due to limited or unreliable information about the assignment location.

Overseas labor costs

Some cultures, by their inherent nature, offer lower labor costs for international operations. Consequently, the search for a competitive operating site might take an organization to a culture that is totally different from the base station. The pervasiveness of fast routing of information makes it easy for a business to find a seemingly receptive overseas site to relocate operations. The consequence of such relocation is a transfer of culture in either direction, often with limited situational awareness. Many organizations don't fully appreciate the differences in the operating cultures. They pay attention only to the physical and economic aspects of their operations. But more often than not, the cultural shock and unsuccessful assimilation (from either end of the culture transfer) can lead to project failures.

Many of the economically underserved countries in Asia, Africa, Oceania, and Latin America are frequent targets to international project development. Those countries have common characteristics, such as highly dependent economies (devoted to producing primary products for the developed world), traditional and rural social structures, high population growth, and widespread poverty. Certain characteristics exist that may constitute barriers to successful global projects. Some of these are as follows:

- Limited access to information (substandard telecommunications infrastructure)
- Politically induced trade barriers
- Cultural norms that impede free flow of information
- Existence and abundance supply of cheap, albeit untrained, workforce
- Orientation of manpower toward artisan and apprenticeship labor

Awareness of overseas workforce constraints

Most project outsourcing points are located in developing and underdeveloped nations. These locations often have repressive cultures that are replete with norms that the Western world would find unacceptable. A cultural bridge usually is missing between the developed nations and the developing nations with respect to workforce capabilities. Thus, there are increasing cultural and economic disparities between global business partners. Some of the local issues to be factored into global situational awareness programs include the following:

- Poverty
- Pollution
- Disease
- Inferior health services
- Political oppression
- Gender biases
- Economic and financial scams
- Wealth inequities
- Social permissiveness among the elite

Project personnel posted to these regions are shocked by the level of cultural differences that they experience. In some cases, they maintain a *laissez faire* and hands-off attitude. But there have also been cases where some of the personnel take advantage of the loose culturally acceptable social contacts that could act to the detriment of an international project.

Hierarchy of needs in a different culture

The psychology theory of *hierarchy of needs* proposed by Abraham Maslow (1943) in his paper, "A Theory of Human Motivation," still governs how different cultures respond along the dimensions of global project expectations. A culturally induced disparity in the hierarchy of needs implies that we may not be able to fulfill our project responsibilities along the global spectrum of international projects. In a culturally different workforce, the specific levels and structure of the needs may be drastically different from the typical mode observed in the Western culture. Maslow's hierarchy of needs consists of five stages:

1. *Physiological needs*: These are the needs for the basic necessities of life, such as food, water, housing, and clothing (i.e., survival needs). This is the level where access to money is most critical.
2. *Safety needs*: These are the needs for security, stability, and freedom from physical harm (i.e., desire for a safe environment).
3. *Social needs*: These are the needs for social approval, friends, love, affection, and association (i.e., desire to belong). For example, social belonging may bring about better economic outlook that may enable each individual to be in a better position to meet his or her social needs.
4. *Esteem needs*: These are the needs for accomplishment, respect, recognition, attention, and appreciation (i.e., desire to be known).
5. *Self-actualization needs*: These are the needs for self-fulfillment and self-improvement (i.e., desire to arrive). This represents the stage of opportunity to grow professionally and be in a position to selflessly help others.

In an economically underserved culture, most workers will be at the basic level of physiological needs, and there may be cultural constraints on moving from one level to the next higher level. This fact has an implication on how cultural interfaces may fail between host and guest nations involved in a global project.

Project sustainability

Sustainability—it isn't just for the environment. Project sustainability on the global front is as much a need as the traditional components of project management spanning planning, organizing, scheduling, and control. For international projects, preemption of cultural conflicts is far better than post-occurrence remedies. Proactive pursuit of project best practices can pave the way for project success on a global scale.

Conclusion

Cultural infeasibility is one of the major impediments to project outsourcing in an emerging economy. The business climate of today is very volatile. This volatility, coupled with cultural limitations, creates problematic operational elements for global projects located in a developing country.

For example, an inquiry or revelation of personal information is viewed as taboo in many developing countries. Consequently, this may impede the collection, storage, and distribution of workforce information that may be vital to the success of global project management. Global situational awareness and its best practices are essential for any organization engaged in international projects. In these days of globalization, nothing happens without international cooperation, and cooperation cannot succeed without effective situational awareness on each side of the cooperating partners.

References

Badiru, A. B., B. O. Nnaji, G. Okogbaa, P. Egbelu, *Proceedings of the Sixth Africa-USA International Conference on Manufacturing Technology*, MANUTECH 2002, Abuja, Nigeria, July 8–11, 2002.

Badiru, A. B., B. O. Nnaji, G. Okogbaa, P. Egbelu, *Proceedings of the Seventh Africa-USA International Conference on Manufacturing Technology*, MANUTECH 2004, Port Harcourt, Nigeria, July 12–14, 2004.

Bhargava, V., editor, *Global Issues for Global Citizens: An Introduction to Key Development Challenges*, The World Bank, Washington, DC, 2006.

Ibidapo-Obe, O., Models for technology appropriate for sustainable development, in B. O. Nnaji et al., editors, *Proceedings of the Third Africa-USA International Conference on Manufacturing Technology*, Accra, MANUTECH 1996, Ghana, August 12–15, 1996, pp. 193–194.

Maslow, Abraham H., A theory of human motivation, *Psychology Review*, 1(3), 1943, pp. 93–100.

Nnaji, B. O., A. B. Badiru, G. Okogbaa, P. Egbelu, *Proceedings of the First Africa-USA International Conference on Manufacturing Technology*, MANUTECH 1993, Ikeja, Lagos, Nigeria, January 11–14, 1993.

Nnaji, B. O., A. B. Badiru, G. Okogbaa, P. Egbelu, *Proceedings of the Second Africa-USA International Conference on Manufacturing Technology*, MANUTECH 1994, Ikeja, Lagos, Nigeria, August 8–11, 1994.

Nnaji, B. O., A. B. Badiru, G. Okogbaa, P. Egbelu, *Proceedings of the Third Africa-USA International Conference on Manufacturing Technology*, MANUTECH 1996, Accra, Ghana, August 12–15, 1996.

Nnaji, B. O., A. B. Badiru, G. Okogbaa, P. Egbelu, *Proceedings of the Fourth Africa-USA International Conference on Manufacturing Technology*, MANUTECH 1997, University of Pittsburgh, Pittsburgh, PA, August 11–14, 1997.

Nnaji, B. O., A. B. Badiru, G. Okogbaa, P. Egbelu, *Proceedings of the Fifth Africa-USA International Conference on Manufacturing Technology*, MANUTECH 2000, Abuja, Nigeria, July 10–14, 2000.

Okogbaa, G., A. B. Badiru, B. O. Nnaji, P. Egbelu, *Planning Document for the Eighth Africa-USA International Conference on Manufacturing Technology*, MANUTECH 2007, Knoxville, TN, July 18–20, University of Tennessee, Knoxville, TN, 2007.

chapter two

Industrialization through manufacturing

Manufacturing is a key foundation for national development. Industrialization is one of the primary means of improving the standard of living in a nation. Manufacturing provides the foundation for sustainable industrialization. Manufacturing refers to the activities and processes geared toward the production of consumer products. Industrialization refers to the broad collection of activities, services, resources, and infrastructure needed to support business and industry. History indicates the profound effect that the Industrial Revolution of the nineteenth century had on world development. Industrialization will continue to play an active role in national economic strategies. Nations that continue to seek international aid for one thing or another can lay a solid foundation for industrialization through manufacturing enterprises. A nation that cannot institute and sustain manufacturing will be politically delinquent and economically retarded in the long run. A good industrial foundation can positively drive the political and economic processes in any nation.

In order to achieve and sustain manufacturing development, both the technical and managerial aspects must come into play. This book contains the managerial processes necessary to facilitate manufacturing development. Project management and technology management are presented as a viable means of achieving manufacturing development. Technology transfer is a key component of technology management.

Complexity of manufacturing

The economic and social problems facing many nations in the present technological age are very challenging. The gloomy trends have been characterized by stagnant state of established industries, decline in productivity, closure of poorly managed corporations, globalization of markets, increased dependency on physical technology, increased apathy toward social issues, proliferation of organized and sophisticated illegal financial deals, requirement for expensive capital investments, and neglect of economic diversification endeavors. All these problems led to the decline of manufacturing in the 2000s. Fortunately, a resurgence of manufacturing is presently being experienced in different parts of the world.

An industrial development project is a complex undertaking that crosses several fields of endeavors. The diverse political, social, cultural, technical, organizational, and economic issues that intermingle in industrial development compound manufacturing efforts. Sophisticated managerial approaches are needed to control the interaction effects of these issues. Financial power is a necessary but not a sufficient requirement for industrial development. A common mistake by a developing economy is to simply dump scarce financial resources at a development problem without making adequate improvement in the management processes needed to support the development goal. Focusing on the technical aspects of a development project to the detriment of the managerial aspects will only create the potential for failure. A project management approach can facilitate an integrated understanding of the complex issues involved in industrial development and, thus, pave the way for success.

Contrary to general belief, industrial development problems are not limited to the developing nations alone. Even in developed and industrialized nations, large pockets of industrially neglected communities can be found. Residents of these communities live in abject poverty despite the ambient affluence of the overall nation. The countries that are most blessed with natural resources are often those that suffer most from industrial neglect. The problem is typically that they do not know how to initiate and implement the projects that are needed to exploit the available resources. Some rely on misaligned technology transfers. As portrayed in Figure 2.1, technology aligned is essential for achieving sustainable development goals. If technology fidelity is high, but alignment is low, then the effectiveness of the transferred will be low (lower left region). If technology fidelity is high and alignment is high, then we have the greatest potential for success (upper right region).

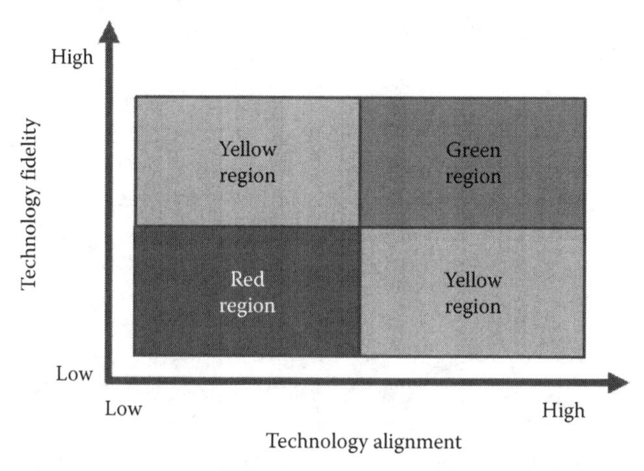

Figure 2.1 Technology transfer fidelity and alignment.

Proper applications of project management techniques can facilitate further responsiveness and effectiveness to pave the way for the advancement of manufacturing.

Strategic planning for manufacturing

Industrial planning determines the nature of actions and responsibilities required to achieve industrial development goals. Strategic planning involves the long-range aspects of manufacturing development efforts. Planning forms the basis for all actions. Strategic planning for manufacturing development can be addressed at three distinct levels, as discussed below.

Supra-level planning

Planning at this level deals with the big picture of how manufacturing development fits the overall and long-range needs of the community, the region, or nation. Questions faced at this level may concern the potential contributions of manufacturing activities to the standard of living in the community, the development of resources needed to provide basic amenities in the community, the required interfaces between development projects within and outside the community, the government support for manufacturing development, responsiveness of the local culture, and political stability.

Macro-level planning

Planning at this level may address the overall planning within a defined industrial boundary. The scope of the development effort and its operational interfaces should be addressed at the macro-level planning. Questions addressed at this level may include industry identification, product definition, project scope, availability of technical manpower, availability of supporting resources, workforce availability, import/export procedures, development policies, effects on residential neighborhoods, project funding and finances, and project coordination strategies.

Micro-level planning

This level of planning deals with detailed operational plans at the task levels of manufacturing activities. Definite and explicit tactics for accomplishing specific development objectives should be developed at the micro level. Factors to be considered at the micro-level planning may include scheduled time, training requirement, tools required, task procedures, reporting requirements, and quality requirements. Manufacturing is capital intensive. If not planned properly, the investment may prove to be an uneconomic venture.

Industrial development and economic development

Industrialization is one side of the economic coin. Industrial development can directly translate to economic development if proper management practices are followed. Industrial development can be formulated as a basic foundation for economic vitality and national productivity improvement. Some of the major factors that can positively impact interactions of industrial and economic processes include the following:

- Unification of national priorities
- Diversification of the economic strategy
- Strong strategy for development of rural areas
- Adequate investment in research and development
- Stable infrastructure to support development effort
- Political stability that prevents economic disruption
- Social and cultural standards that improve productivity

Industrial development plays different types of roles in economic development. The structure of certain nations is such that the economic and industrial systems are on different development tracks. To facilitate the interaction of industrial and economic endeavors, plans must be made to allow the industrial system to coexist symbiotically with other production forces.

Pursuit of technological change

Technological change is needed to drive industrial development. Technological progress calls for the use of technology. Technology, in the form of information, equipment, and knowledge, can be used productively and effectively to lower production costs, improve service, improve product quality, and generate higher output levels. The information and knowledge involved in technological progress include those which improve the performance of management, labor, and raw materials.

Technological progress plays a vital role in improving total productivity. Statistics on developed countries, such as in the United States, show that in the period 1870–1957, 90% of the rise in real output per man-hour can be attributed to technical progress. It has been shown that industrial or economic growth is dependent on improvements in technical capabilities as well as on increases in the amount of the conventional factors of capital and labor. Technological change is not necessarily defined by a move toward the most modern capital equipment. Rather, technological changes should be designed to occur through improvements in the efficiency in the use of existing equipment. The challenge to developing countries, such as those in Africa, is how to develop the infrastructure that promotes and utilizes available technological resources.

Social change and development

Industrial development requires change. A society must be prepared for change in order to take advantage of new industrial manufacturing opportunities. Efforts that support industrial development must be instituted into every aspect of everything that the society does. If a society is better prepared for change, then positive changes can be achieved. Industrial development requires an increasingly larger domestic market. The social systems that make up such markets must be carefully coordinated. The socioeconomic impact of industrialization cannot be overlooked. Social changes are necessary to support industrial development efforts. Social discipline and dedication must be instilled in the society to make industrial changes possible. The roles of the members of a society in terms of being responsible consumers and producers of industrial products must be outlined. People must be convinced of the importance of the contribution of each individual, whether that individual is acting as a consumer or as a producer. Industrial consumers have become so choosy that they no longer will simply accept whatever is offered in the marketplace. In cases where social dictum directs consumers to behave in ways not conducive to industrial development, changes must be instituted. If necessary, acquired taste must be developed to like and accept the products of local industry.

To facilitate consumer acceptance, the quality of domestic industrial products must be improved to competitive standards. New technologies that facilitate increased industrial productivity must be embraced whether through domestic sources or through technology transfer. For an industrial product to satisfy the sophisticated taste of the modern consumer, it must exhibit a high level of quality and responsiveness to the needs of the consumer. Only high-quality industrial products and services can survive the increasing competitive market. Some of the approaches for preparing a society for industrial development changes are listed as follows:

- Highlight the benefits of industrial development
- Keep citizens informed of the impending changes
- Get citizen groups involved in the decision process
- Promote industrial change as a transition to a better society
- Allay the fears about potential loss of jobs due to industrial automation
- Emphasize the job opportunities to be created by industrial development

Stipulation for the use of local raw materials is one policy that may necessitate a change in social attitudes. The society (consumers and producers) must understand the importance of local raw materials in domestic manufacturing activities. When backed against the wall of production, producers must develop the necessary indigenous technology or processes for using local materials. The technology transfer modes presented in this book provide a guideline for local adaptation of inwardly transferred technology.

Education of producers and consumers

A more aware population is a more responsive and supportive population. It is essential to include education, training, and public awareness into the strategic planning for manufacturing development. For example, inflation can adversely affect investments in manufacturing development. Inflation is a disease that feeds on itself, but not without the help of people. The basic attitude of citizens toward consumption and wealth greatly determines the trend of inflation. Everyone wants to buy and sell consumer products without paying attention to the production efforts needed to generate those products. The seller wants to sell at whatever exorbitant prices he or she can get. The naive buyer does not help the situation either. He or she is willing to buy whatever he or she wants at whatever the asking price even if it means sacrificing economic personal economic sense, which, over the long run, with many like-minded consumers, can jeopardize the economic path of the whole nation.

The producer, encouraged by the high demand, shrewdly cuts production and, thus, reduces market supply, which then elevates consumer prices even further. The fact that a product is in short supply makes the consumer yearn for more of it. The product quickly becomes a status symbol. The few who can get it claim to be the "class" of the society, and they are willing to pay whatever price to maintain that status quo. Each consumptive individual is unaware of the economic agony that reckless personal acts can cause for the larger population. It is not that consumers are unpatriotic. Rather, it is just that they don't know the adverse implications of collective personal economic habits. This is why educating the public is an important aspect of combating inflation and paving the way for manufacturing development. The producer, the retailer, and the buyer all need to know the real causes of inflation, its boomerang effects, and its potential remedies. The government should play a very active role in the education process. The indoctrination process needs to start right at the grassroots. Children should be exposed to the concepts of battling inflation so that they can grow up to become responsible producers, retailers, or consumers. The early instillation of these concepts in the youths will even have the beneficial side effect of imbibed national pride, which should serve the social and political interests of the nation in later years. Most of the leadership problems in many countries can be traced to a lack of deep-rooted pride and interest in the national welfare.

In an underdeveloped nation, nothing is more discouraging than to hear people proclaim how they have given up on their nation. They claim that nothing works, that no improvements are forthcoming, and that the government is unresponsive to the suffering of the people. How do we expect citizens who feel negative about their nation perform when they, themselves, reach the position of national leadership? Certainly, attitudes and perception will not change overnight. If citizens carry negative impressions into higher offices, all they will do is to perpetrate their views by saying, "Didn't I say nothing works in this country?" Obviously, such people have not had early mechanisms of instilling national pride and economic dedication in them. This is why early education and public awareness of economic processes are essential for creating sustainable manufacturing activities.

Government programs cannot succeed without the people's cooperation in the implementation process. Herein lies the irony of governing and being governed. We blame the government for not instituting corrective programs and we turn right around to take actions that innocently sabotage the prevailing programs, thereby compounding the burden of the government. People must be educated on the implications of their actions. Presenting a national strategy is not enough. Ensuring that citizens are educated about how to support and execute the strategy is equally essential. When a consumer product is in short supply, everyone complains and blames the system. No one reflects on the individual contribution at each consumer or producer level. Consider the case of a factory worker who buys a locally made product and complains of its inferior quality compared to imported brands. He or she forgets the personal unique position to contribute significantly to the quality of local products through on-the-job activities. There are many cases where auto industry workers buying foreign-made vehicles not because of the need for diversity of assets, but rather due to the belief of the better quality of imported vehicles. This is like sending a signal that says "I build the local vehicle and I know about it. Please don't buy it!" That is a fallacy of production dedication. The worker bites the hand that provides the feeding.

To further illustrate the problem of lost dedication, consider the hypothetical scenario of an auto industry employee. As the demand for imported vehicles increases so do the prices of the vehicles. So, the employee asks the employer for higher wages so that a higher-end imported vehicle can be purchased. In order to pay those higher prices, the employer raises the prices of the locally made vehicles to make more money to pay the higher wages. When the prices of the local vehicles go up without a comparable increase in quality (thanks to the employee's work delinquency), the public buys even less of the locally made vehicles. So, the employer sells fewer cars and makes less money. In the end, the employee is laid off. This shows that the employee's economic education is lacking in the first place. This is why it is essential to broaden the sense of economic awareness of consumers.

The scenarios painted here demonstrate the fact that the economic system needed to support manufacturing development is very complex. Individual actions, as innocent and minor as they may appear, can go a long way in advancing or impeding the overall economy. We should be convinced that the education of the public is a key factor in solving many economic problems before industrial development can take root.

Political stability and manufacturing development

Politics determines the crux of economic vitality and economic vitality determines the crux of politics. With the changing political environments in developing nations, as those in Africa, it is difficult to stabilize government policies. Many governments are perpetually in transient states. Just when policies are beginning to take a firm hold, a new administration comes in and, as a show of new power, overturns all previous achievements. Billions of dollars have been wasted by succeeding governments through the absurd practice of abandoning projects started by preceding administrations regardless of merit. For the sake of national progress, governments should make a commitment to retain and execute worthwhile projects irrespective of who started them. Foresight should also be exercised when embarking on new industrial projects to ensure that subsequent governments can find the merit for their continuation. Laying a solid foundation for an industrial development project can facilitate a lasting coexistence of political, economic, and industrial activities. The major role of government in industrial development must be that of a facilitator. This is particularly important in intergovernment negotiations for mutual development. The interdependence of the political, economic, and industrial systems is depicted in Figure 2.2. It is noted that manufacturing vitality advances along the manufacturing system axis depending on the advances in the economic and political systems. The building blocks of industrial performance are represented by the industrial performance surface and are dependent on the positive interplay of manufacturing, economic, and political systems.

If the economic and political processes are at low levels, no significant industrial development can be expected. As the economy picks up and the political atmosphere improves, the level of industrial development will begin to increase. To maximize industrial output, the political and economic systems must operate at symbiotically high levels. This is why a developing nation must recognize that chaotic political situations will adversely affect overall national development efforts.

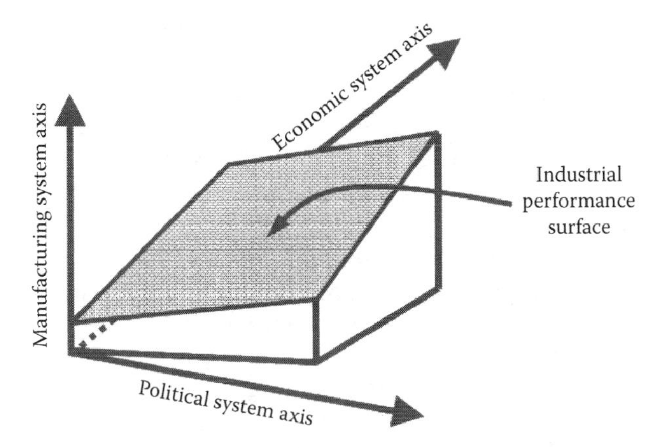

Figure 2.2 Interdependence of political, economic, and industrial systems.

Global influences on domestic manufacturing

The new world economic order calls for new approaches to development. The prevailing changes around the world will have profound effects on international trade and domestic manufacturing activities. The changes in eastern Europe, the advancements in western Europe, the merging of nations, the breakup of nations, and the emergence of Africa as a viable market will all affect international trade in the coming years. No nation can be insulated from what goes on in other nations. If not impacted directly, impacts can occur through intermediary trading partners. The mobility of manufacturing outputs and products will be one common basis for global trade communication. Companies and countries must recognize the trend and refocus efforts and direct investments accordingly. Some of the key aspects of globalization that impact industrial development activities include the following:

- Transition of some countries from being trade partners to being trade competitors
- Reduction in production cycle time to keep up with the multilateral introduction of new products around the world
- Increased efforts to reduce product life cycle from years to months
- Increased responsiveness to the needs of a mixed workforce
- The pressure to mitigate the adverse impacts of cultural barriers
- Relaxation and expansion of trade boundaries
- Tightening of trade relationships
- Politically driven trade sanctions
- Integration of operations and services and consolidation of efforts
- More effective and responsive communication, including social media

- Increased pressure for multinational cooperation
- Need for multicompany and multiproduct coordination
- National security barriers

The interdependence of world projects in modern times makes it imperative that global development efforts be pursued. Political, industrial, economic, and social disasters in one nation can easily spill over to other nations and create adverse chain reactions. Refugee problems now plaguing many nations are cruel reminders that nations can no longer exist in isolation.

There are several unique factors that impinge on industrialization. Some of the factors are addressed in the sections that follow. While technology continues to push the limit of perfection in developed nations, citizens of underdeveloped nations continue to be relegated to antiquated production tools. Industrial development has unique aspects that must be considered in formulating development strategies. Some of these aspects are discussed below.

Rural area development for industrialization

Because of their underdeveloped state, rural areas offer significant potentials for industrial development. Ironically, it is that same underdevelopment that makes them susceptible to neglect. Productive members of rural areas move to urban areas in large numbers in search of a better life. The more the technological advancement in a nation, the more the rural residents migrate to urban areas. Consequently, the manpower needed to support the development of a rural area is often not consistently available. The key to keeping the manpower in their localities is the development of the localities and the creation of employment opportunities and the provision of basic amenities of life. Rural neglect has created highly uneven distribution of population in most countries. Urbanization without industrialization is doomed to economic failure.

The sparse population in rural areas means large land areas with potential for industrial setups are available at relatively low cost. This makes them particularly attractive for the location of new plants. But investors often shy away from the rural areas because of the limited availability of supporting infrastructure such as electrical power, water, communication, raw materials, and transportation facilities. If power, water, transportation, and communication are guaranteed, the development of the rural areas will be effected successfully. In addition to assuring these services, a nation should offer incentives (e.g., tax incentives) to rural investors. Some key elements of rural area development include the following:

- Improvement of basic infrastructure
- Creation of investment incentives for local businesses

- Establishment of academic institutions geared to local educational needs
- Establishment of technology extension services through academic institutions
- Creation of a local clearing house for technology information
- Government assistance for local product development
- Provision of adequate rural healthcare services

Cultural barriers to industrialization

Cultural barriers can inhibit industrial development. Most industrialization programs are fueled by foreign ideas, technologies, and approaches. In some cultures, cloaks of protection have been developed to protect the society against foreign influences. The greatest fear often involves the potential moral decadence that may accompany industrial development as people migrate from one nation to another in response to the development process. Cultural barriers may also prohibit certain segments of a society from participating fully in the development process. Such cultural barriers must be removed to pave the way for full participative industrial development. Some cultures have evolved to a level of indifference to fraudulent and corrupt practices that are detrimental to development efforts. In some cases, corruption is viewed as an accepted way of doing business. Culturally permissive attitudes to corruption must be altered before general development can take hold. Loyalty to the overall development goals rather than individual pursuits is required as a part of cultural changes for industrial development.

Education and industrial development

He who owns the knowledge controls the power. Both formal and informal education should play a vital role in national industrial development. In the modern day of high technology, adequate education is needed to succeed in any work environment. Even in some industrialized nations, a large percentage of the adult population in neglected communities is functionally illiterate. It used to be that the children of these poverty-stricken communities were needed to drive the wheels of manual labor in local factories. But the present and future industries, with increasing push for automation, will not need much of the services of the labor-intensive workforce. Education geared toward the new industrial direction will be needed to participate actively in industrial development. Poor parents always proclaim that they don't want their children to end up where they did. Yet, no drastic educational efforts are made to ensure that they don't.

Research and development partnership

Research and education go hand in hand in creating pathways for industrial development. Partnerships between government, university, and industry offer good avenues for addressing the pressing issues of national development. Figure 2.3 illustrates a multifaceted approach to government–university–industry partnerships. There is enough room for each entity's goals. On the side of business and industry, the end goal is to maximize profit and ensure survival. On the side of the university, the primary goal is the quest for new knowledge, technology advancement, and technology transfer. On the part of the government, the goal is to increase the wealth and security of the nation. Specific research-based partnerships can address the multilateral goals. Quantitative tools such as multiattribute decision modeling are of interest for this purpose.

Knowledge is a sustainable capital. The establishment of a formal process for the interface of institutions of higher learning and industry can be one of the capitals for industrial development. Universities have unique capabilities that can be aligned with industry capabilities to produce symbiotic working relationships. Private industrial research projects must complement public industrial research programs. Academic institutions

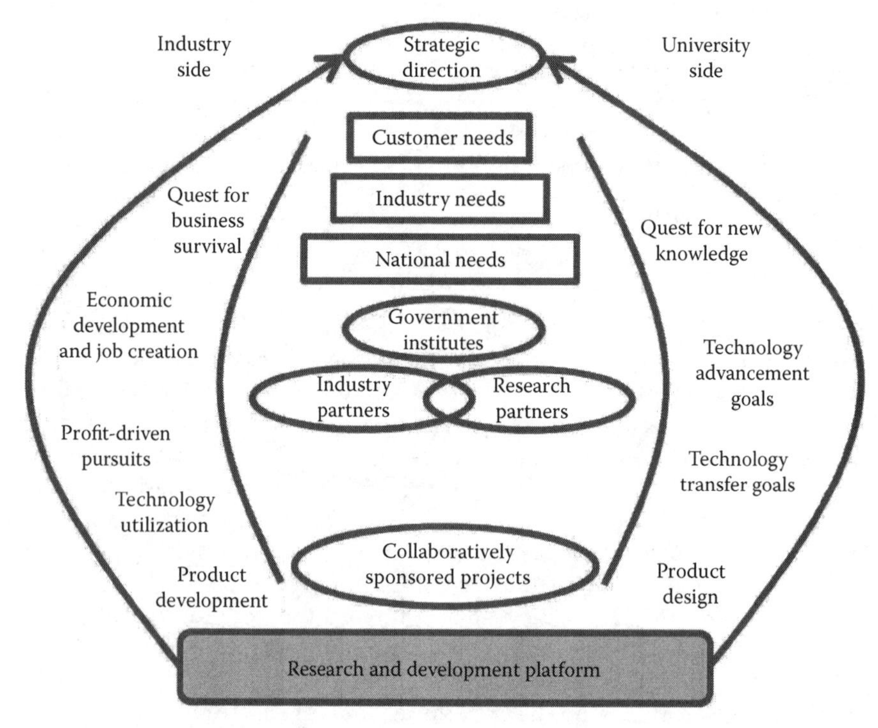

Figure 2.3 Multitiered partnership of government, industry, and university.

have a unique capability to generate, learn, and transfer technology to industry. The quest for knowledge in academia can fuel the search for innovative solutions to specific industrial problems. A collaborative industry is a fertile ground for developing prototypes of new academic ideas. Industrial settings are good avenues for practical implementation of technology. Industry-based implementation of university-developed technology can serve as the impetus for further efforts to develop new technology. Technologies that are developed within the academic community mainly for research purposes often languish in the laboratory because of the lack of formal and coordinated mechanisms for practical industrial implementation. The potentials of these technologies go untapped for several reasons including the following:

- The researcher does not know which industry may need the technology.
- Industry is not aware of the technology available in academic institutions.
- There is no coordinated mechanism for technical interface between industry and university groups.

Universities interested in technology development are often hampered by the lack of adequate resources for research and training activities. Industry can help in this regard by providing direct support for industrial groups to address specific industrial problems. The universities also need real problems to work on as projects or case studies. Industry can provide these under a cooperative arrangement. The respective needs and capabilities of universities and industrial establishments can be integrated symbiotically to provide benefits for each group. University courses offered at convenient times for industry employees can create opportunity for university–industry interaction. Class projects for industry employees can be designed to address real-life industrial problems. This will help industry employees to have focused and rewarding projects. Class projects developed in the academic environment can be successfully implemented in actual work environments to provide tangible benefits. With a mutually cooperative interaction, new developments in industry can be brought to the attention of academia while new academic research developments can be tested in industrial settings. The growing proliferation of distance learning and distance education programs can be leveraged to spread educational opportunities to more industry.

The role of government should be as the facilitator to provide linkages and programs that enable multiorganizational collaboration. One desirable approach is for the government to establish and fund technology clearing houses. Academic institutions can serve as convenient locations for technology clearing houses. Such clearing houses can be organized to provide up-to-date information for industrial development activities. Specific industrial problems can be studied at the clearing house. The clearing house can serve as a repository for information on various technology tools. Industry would participate in the clearing house through the donation of equipment, funds, and personnel time. The services provided by a clearing house could include a combination of the following:

- Provide consulting services on technology to industry
- Conduct on-site short courses with practical projects for industry
- Serve as a technology library for general information
- Facilitate technology transfer by helping industry identify which technology is appropriate for which problems
- Provide technology management guidelines that will enable industry to successfully implement new technology in existing operations
- Expand training opportunities for engineering students and working engineers

The establishment of centers of excellence for pursuing industrial development related research is another approach to creating a favorable atmosphere for industry–university interaction. As an example, from 2000 to 2006, the author founded and directed the Industrial Engineering Center for Industrial Development Research within the Department of Industrial Engineering at the University of Tennessee in Knoxville, Tennessee. The center was dedicated to research studies involving the multidimensionality of factors and processes affecting the movement of a product from concept stage to the commercialization stage with specific ties to regional economic development. A part of the services of the center was industry location feasibility studies.

Agriculture and industrialization

A hungry society cannot be an industrially productive society. It is generally believed that an underdeveloped economy is characterized by an agricultural base. Based on this erroneous belief, several developing nations have abandoned their previously solid agricultural base in favor of alternate means of industrialization. The fact is that a strong agricultural base is needed to complement other industrialization efforts. Agriculture, itself, is a viable source of industrialization when we consider the broad definition of industrialization presented Chapter 2. Mechanized farming is a sort of industrialization. If industrialization does not yield immediate benefits, the society will be exposed to the double jeopardy of hunger and material deprivation. Once abandoned, agriculture is a difficult process to regain. Since agricultural processes take several decades to perfect, revitalization of abandoned agriculture may require several decades.

Agriculture should play a major role in the foundation for industrial development. The agricultural sector can serve as a viable market for a concomitant industry through supply chain interfaces. It is interesting to note that the agricultural revolutions of the past pave the way for the subsequent industrial revolution.

Human history indicates that humans started out as nomad hunters and gatherers, drifting to wherever food could be found. About 12,000 years ago, humans learned to domesticate both plants and animals. This agricultural breakthrough allowed humans to become settlers, thereby spending less time wandering in search of food. More time was, thus, available for pursuing stable and innovative activities, which led to discoveries of better ways of planting and raising animals for food. That initial agricultural discovery eventually paved the way for the agricultural revolution. During the agricultural revolution, mechanical devices, techniques, and storage mechanisms were developed to aid the process of agriculture. These inventions made it possible for more food to be produced by fewer people. The abundance of food meant that more members of the community could spend that time for other pursuits rather than the customary labor-intensive agriculture. Naturally, these other pursuits involved the development and improvement of the tools of agriculture. The extra free time brought on by more efficient agriculture was, thus, used to bring about more technological improvements in agricultural implements. These more advanced agricultural tools led to even more efficient agriculture. The transformation from the digging stick to the metal hoe is a good example of the raw technological innovation of that time. With each technological advance, less time was required for agriculture, thereby, permitting more time for further technological advancements. The advancements in agriculture slowly led to more stable settlement patterns. These patterns led to the emergence of towns and cities. With central settlements away from farmlands, there developed a need for transforming agricultural technology to domicile technology that would support the new city life. The transformed technology was later turned to other productive uses which eventually led to the emergence of the industrial revolution. To this day, the entwined relationships between agriculture and industry can still be seen. Figure 2.4 shows a representation of the intersection of manufacturing technologies.

Workforce for industrialization

Technical, administrative, and service workforce will be needed to support industrialization initiatives. People make development possible. No matter how technically capable a machine may be, people will still be required to operate or maintain it. Soon after World War II, it was generally

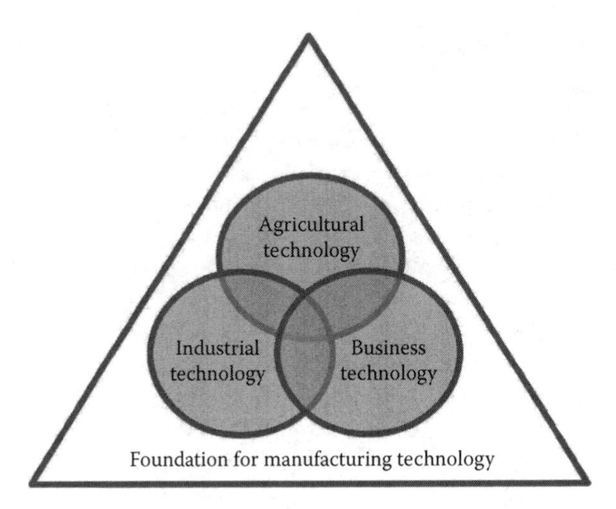

Figure 2.4 Intersection of manufacturing technologies.

believed that physical capital formation was a sufficient basis for development. That view was probably justified at that time because of the role that machinery played during the war. It was not obvious then that machines without a trained and skillful workforce did not constitute a solid basis for development. It has now been realized that human capital is as crucial to development as physical capital. The investment in workforce development must be given a high priority in the development plan. Some of the important aspects of manpower supply analysis for industrial development include the following:

- Level of skills required
- Mobility of the manpower
- The nature and type of skills required
- Retention strategies to reduce brain drain
- Potential for coexistence of people and technology
- Continuing education to facilitate adaptability to technology changes

Who is available to educate and train the workforce

With respect to who may be available to contribute to the education and training of the national workforce, the author offers the following perspectives. Military service has an impact in advancing workforce education. By virtue of having gone through structured education and training for their service requirements, ex-servicemen/women do have unique skills and experience that can be channeled toward educating and training the national workforce for specific needs. Military education can have far more impact on national advancement than we realize. In the days of the author's own university studies in the United States, he marveled at the knowledge base, span of expertise, and professionalism of many of the engineering instructors. He wondered how such a high concentration of marvelous militarily experienced engineers could be found in one institution. He later found out that this was not an isolated incident in one institution. It turned out that a large number of engineering professors in the 1960s and 1970s across the United States had served in the U.S. Navy or Army during World War II. After the war, through government programs, many transferred their military training, education, and expertise into lecturing at universities. The positive impression that the author had of his engineering professors gave him the early incentives to apply more forthright efforts to his engineering education and, subsequently, chose academia as his own career path. The consequence is that the foundational knowledge acquired from the military-engineers-turned professors continues to serve the author's own students years later. The conclusion is that the military directly and indirectly influenced the advancement of technical workforce in the United States. What the United States is experiencing today in terms of being a world leader is predicated on a foundation of consistent technical education over the years. Other nations, particularly developing nations, can learn from this example.

There is often a debate whether the military should continue to invest (and how much) in advanced military education. The fact is that the national investment in advanced military education is not only essential to keep the military on the cutting edge of warfare technology, but also to positively impact the national landscape of education that may contribute to the advancement of manufacturing technology. Recognizing the urgent need to address global societal issues from a technical standpoint, in 2008, the National Academy of Engineering (NAE 2008) of the United States published the *14 Grand Challenges for Engineering*. The challenges have global implications for everyone, not just the engineering professions. As such, solution strategies must embrace all disciplines. The military, by virtue of its global presence and wider span of involvement in advanced technical education, can provide the technical foundation for addressing many of the challenges. Engineers of the future will need diverse skills to tackle the multitude of issues and factors involved in adequately and successfully addressing the challenges. Military engineers, in particular, are needed to provide the diverse array of technical expertise, discipline, and professionalism required. Technical education, covering science, technology, engineering, and mathematics at advanced levels, provides a sustainable opportunity for the military to impact the 14 grand challenges listed as follows:

1. Make solar energy economical
2. Provide energy from fusion
3. Develop carbon sequestration methods
4. Manage the nitrogen cycle
5. Provide access to clean water
6. Restore and improve urban infrastructure
7. Advance health informatics
8. Engineer better medicines
9. Reverse-engineer the brain
10. Prevent nuclear terror
11. Secure cyberspace
12. Enhance virtual reality
13. Advance personalized learning
14. Engineer the tools of scientific discovery

All of the above have implementation implications for manufacturing technology, either directly or indirectly through transitional interactions. An extract from the National Academy of Engineering (2008) document on the 14 grand challenges reads: "In sum, governmental and institutional, political and economic, and personal and social barriers will repeatedly arise to impede the pursuit of solutions to problems. As they have throughout history, engineers will have to integrate their methods and solutions with the goals and desires of all society's members." Who is better capable to analyze, synthesize, and integrate than the technically trained military? Advanced military education is needed and should be sustained to carry out this charge. A recommendation is that qualified ex-service people should be sent back to school for advanced technical education and turn them into future engineering educators so that future engineering students can benefit and the nation as a whole can also benefit. This may even have the side benefit of having the society view the military in the positive light of national economic advancement beyond national security.

Technology attributes for industrialization

Some specific attributes of technology can facilitate industrial development. But technology must be managed properly to play an effective role. There is a multitude of new technologies that has emerged in recent years. Hard and soft technologies such as computing tools, cellular manufacturing, intelligent software tools, and social media are changing the landscape of manufacturing rapidly. But much more remains to be done in actual implementation. It is important to consider the peculiar characteristics of a new technology before establishing adoption and implementation strategies for applications in industrial development. The justification for the adoption of a new technology should be a combination of several factors rather than a single characteristic of the technology. The important characteristics to consider include productivity improvement, improved product quality, reduction in production cost, flexibility, reliability, and safety.

An integrated evaluation must be performed to ensure that a proposed technology is justified economically and technically. The scope and goals of the proposed technology must be established right from the beginning of an industrialization project. This entails the comparison of industry objectives with the overall national goals in the areas discussed as follows:

- *Market target*: This should identify the customers of the proposed technology. It should also address items such as market cost of the proposed product, assessment of the competition, and the market share.
- *Growth potential*: This should address short-range expectations, long-range expectations, future competitiveness, future capability, and prevailing size and strength of the competition.
- *Impact on national goals*: Any prospective technology must be evaluated in terms of the direct and indirect benefits to be generated by the technology. These may include product price versus value, increase in international trade, improved standard of living, cleaner environment, safer work place, and improved productivity.
- *Profitability*: An analysis of how the technology will contribute to profitability should consider past performance of the technology, incremental benefits of the new technology versus conventional technology, and value added by the new technology.
- *Capital investment*: Comprehensive economic analysis should play a significant role in the technology assessment process. This may cover an evaluation of fixed and sunk costs, cost of obsolescence, maintenance requirements, recurring costs, installation cost, space requirement cost, capital substitution potentials, return on investment, tax implications, cost of capital, and other concurrent projects.
- *Skill and resource requirements*: The utilization of resources (manpower and equipment) in the pre-technology and post-technology phases of industrialization should be assessed. This may be based on material input–output flows, the high value of equipment versus productivity improvement, the required inputs for the technology, the expected output of the technology, and the utilization of technical and nontechnical personnel.

- *Risk exposure*: Uncertainty is a reality in technology adoption efforts. Uncertainty will need to be assessed for the initial investment, return on investment, payback period, public reactions, environmental impact, and volatility of the technology.
- *National productivity improvement*: An analysis of how the technology may contribute to national productivity may be verified by studying industrial throughput, efficiency of production processes, utilization of raw materials, equipment maintenance, absenteeism, learning rate, and design-to-production cycle.

Four-sided infrastructure for industrialization

Industrialization built upon a solid foundation can hardly fail. The major ingredients for durable industrial, economic, and technological developments are electrical power, water, transportation, and communication facilities. These items should have priority in major industrial development projects. Of course, housing infrastructure for the workforce is another basic need that must be accounted for in an industrial development strategy. A four-sided manufacturing infrastructure model to provide a foundation for industrial development is presented in Figure 2.5. Such a development pursuit seeks to ensure the following amenities:

- Primary amenities
  - Reliable power supply
  - Consistent water supply
  - Good transportation system
  - Efficient communication system
- Supporting amenities
  - Housing
  - Education
  - Health care

The provision of adequate healthcare facilities is particularly essential to building a strong industrial base. A healthy society is a productive society while an unhealthy society will be an economically destitute society. Diseases that often ravage impoverished nations can curtail the productive capabilities of the citizens. Recent global problems with contagious diseases, such as the bird flu, SARS, avian flu, enterovirus, and Ebola, have deleterious effects on the productive potentials of the nations affected. The destructive effects of pandemic and epidemic infectious diseases can be mitigated by prompt access to basic healthcare services. With fast global human interfaces nowadays, no nation can be immune to infectious diseases across the continents.

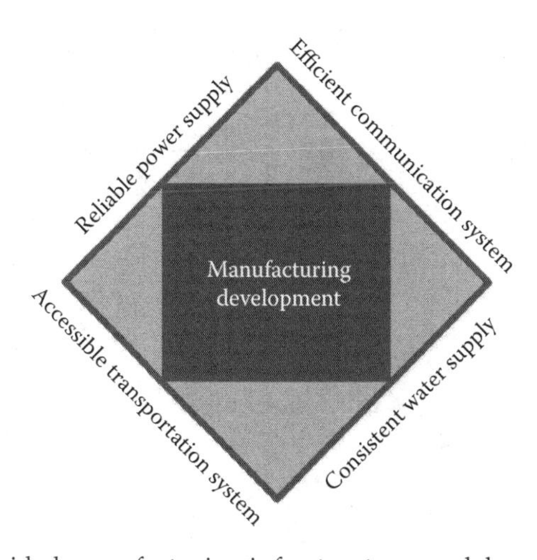

Figure 2.5 Four-sided manufacturing infrastructure model.

Reliable power supply

Electricity is the major source of power for production facilities. This fact has been realized by developing countries for a long time. Yet, not enough effort has been directed at adequate generation and reliable distribution of this very important resource. Where electricity has been generated in abundant quantities, reliable distribution has been miserably lacking. Plants have been shortsightedly constructed without planning for adequate power supply. No wonder then that most industries in developing economies are running only at a fraction of their capacities. The problems plaguing power supply companies should be critically studied so that a lasting solution can be found. A large portion of the initial development efforts should be directed at ensuring adequate power supply. Once supply is found to be reliable and stable, major production endeavors can then be pursued.

Consistent water supply

In today's technology, chemicals play a significant role in product development. Water is an indispensable component of the use of chemicals in industrial operations. Water is needed to not only sustain life, but to also support the many products that make life livable. Water facility development should be addressed in the two categories as follows:

- Industrial water
- Potable water

Because of its huge volume of demand, industry should not compete with households for water supply. Like electricity, reliable water supply should be assured before urging investment in production facilities. There are cases of where a new industry opens, has water supply for a short period, and then is disappointed by the reality of water shortages.

Accessible transportation system

In agreement with Newton's law of motion, nothing moves without transportation. In the ancient days of geographically limited commerce, rapid transportation, though important, was less of a concern. But in the modern society, the global market has necessitated interactions with faraway locations. Developed economies appreciate the necessity of these interactions and they developed transportation systems to meet the needs. Developing economies are "still developing" because their transportation systems, in many respects, are yet to catch up with the rest of the world. Progress in modern commerce depends on the facilities for conveying products from one location to another efficiently, reliably, and safely.

After power and water, transportation should have the third priority in any strategic development plan for manufacturing development. Good roads should be constructed and maintained to provide a lasting support for development efforts. Industrial access roads as well as workforce access roads should be constructed. The economic loss can be enormous if the transportation system is neglected even in a developed economy. For example, The American Society of Civil Engineers reported that 42% of America's major urban highways remain congested, costing the economy an estimated $101 billion in wasted time and fuel annually (ASCE 2013). The report confirms further that the U.S. Federal Highway Administration estimates that $170 billion in capital investment would be needed on an annual basis to significantly improve conditions and performance of the infrastructure. This is a huge financial requirement even for a wealthy nation. For developing nations, this book suggests incremental development, sustainment, and investment in transportation infrastructure. Little blocks of building are more manageable and sustainable. It should be understood that the development of a good transportation system can take a very long time at a very high cost. With its accessibility, the U.S. interstate system has fueled development in several parts of the United States.

Efficient communication system

Communication propels commerce. Telephone, the Internet, and the postal system constitute alternate means of conducting business. In a developing economy, the lack of a reliable telephone service or reliable online systems can force entrepreneurs to make physical appearances when conducting even the most basic of business transactions. This, of course, necessitates the increased use of roads, thereby, overloading the already inadequate transportation system. In an underdeveloped community, business that can be conveniently conducted over the telephone is normally conducted by physical presence. The requirement for physical mobility directly takes a person from productive activities, consequently creating adverse effects on national productivity.

In addition, frequent road transportation increases the exposure of the traveler to the hazards of the precarious transportation system. The more the people need to be on the roads, the higher the congestion, and the greater the risks of road mishaps, which have been known to rob a nation of some of its most productive citizens. The reliability of electric power, water, transportation, and communication services is essential for a sound industrial development. Nations with well-developed communication systems have used them to further their development programs. On the other hand, those who have not paid proper attention to their communication systems will continue to struggle with dormant or regressing economies.

Indigenous economic models

The formulation of an indigenous economic model is relevant for industrial development. The unique aspects of a developing economy that will necessitate indigenous models are discussed in this section. The reasons why imported economic principles and models may not work in a developing nation are presented. The success of indigenous economic models in an underdeveloped area can serve as a useful paradigm for industrial development endeavors in nations with similar economic plights. A synergistic implementation of indigenous economic models will have a positive impact on the overall world economy. Both the economic and noneconomic aspects of industrial development must be accounted for in developing an indigenous model. Well-founded laws and principles of economics have been utilized in the process of formulating existing economic policies. Unfortunately, none of the policies has yielded totally satisfactory results in underdeveloped nations. There has been no short-term success, and there is no guarantee of a long-term success. Restrictions on importation, barter agreements, structural adjustment of currency value, and other economic policies are just a few examples of the concerted efforts being made by some developing countries. The fact that some of these efforts have been or may be unsuccessful indicates that there are certain aspects of the domestic economic situation that are being overlooked, oversimplified, or unrecognized in a developing nation.

The standard economic models that are widely employed throughout the world were developed based on societal behaviors. Many emerging economies, with their peculiarities, simply don't fit the formulation of those models. That, perhaps, is why the economies in those societies have not responded positively to the traditional economic policies. Let us consider the laws of supply and demand for example. The laws are based on the assumption that consumers and producers will respond to certain inputs in some rational fashion. The downward-sloping demand curve indicates that consumers will buy less of a product as the price is increased. By analogy, the upward-sloping supply curve indicates that suppliers will supply more of a product as the price increases. This is quite logical, except in certain underdeveloped economies. In many developing societies, the basic supply and demand laws may not be directly applicable. This is because people are expected to behave in manners foreign to their social, cultural, political, and economic structures and attitudes. Rather than try to mold a developing nation to fit existing economic models, attempts should be made to modify the models to fit the unique situations of the nations. Only then can we have models that accurately explain and predict the actions of *developing* consumers and, thus, provide reliable guidelines for effective national economic policies. Atta (1981) constructs a macroeconomic model of the Ghanaian economy. The model emphasizes the supply side of a small open economy. This presents a good example of an economic view tailored to an indigenous scenario.

The indigenous banking system that has been established in India is another excellent example of an economic strategy tailored to a unique national need. The indigenous banking system is based on the strength of personal relations, which is a strong social link in India. The personal relations are used to vouch for credit worthiness instead of using the conventional collateral approach. This indigenous banking system has worked very well and it has made it possible for rural people, who would have otherwise not gone to the Western-style banks, to actively engage in indigenous banking transactions. In a developing nation, the familiar curves of supply and demand may be mangled by several factors. Some of these factors are quite obvious. Some, on the other hand, are very subtle and can only be fully identified, understood, and quantified through dedicated economic research. Economic research must be directed specifically at formulating indigenous economic models. Some important factors that should be considered in such a research effort include the following:

- The level of propensity to consume in a developing nation (consumption orientation rather than production orientation)
- The inferiority complex that prompts some societies to prefer imported goods and services

- The lack of self-pride in the products of local labor
- The affinity for black-market transactions

The factors mentioned above, in addition to others to be determined through appropriate locally focused research, should form the nucleus of an economic model for a developing country. In 1975, the United Nations initiated a program to ensure that adequate quantitative information is available for the planning of economic and social development in Africa. Today, an indigenous economic model could be a significant component of similar quantitative modeling. It should be recognized that many developing nations are capable of generating their own economic precedents. They typically have the number (in population) and the capability to support indigenous models. Through the practice of hoarding, *developing* retailers already know how to create artificial scarcity. They already know how to mount a social assault on the equilibrium point of the standard supply and demand curves. So, there is nothing like *market price* in existence in many developing nations. Some suppliers have already customized the principles of fair competition to accommodate collusion. Through the forces of collusion, the suppliers can collectively dictate their harsh terms to hapless consumers. In order to lay a serious foundation for industrial development, indigenous economists should be charged with the responsibility of conducting the appropriate research to study the feasibility of indigenous economic models for a nation that is aspiring to be industrialized. This may be effectively accomplished through postgraduate thesis research at the nation's institutions of higher learning. The research may be accomplished under the university–industry cooperative model suggested earlier.

In an in-depth analysis of national development planning, Gharajedaghi (1986) stressed the fact that a plan of action must not be shortsighted by descriptive adjectives that tell us nothing about the nature of the problem being solved. In government functions, there is a tendency to place responsibility for dealing with a problem in that part of government where a relevant functional adjective can be found. For example, if a problem is found in transportation, the blames are directed at the transportation department. The responsibility for dealing with the problem is automatically assumed to be that of the department even though the root of the problem may be a lack of public discipline. Adjectives and nouns used to describe problems (e.g., health, finance, social, economic, and political) often tell us nothing about the real problem being faced. The adjectives simply indicate the point of view of the person looking at the problem. This pitfall must be avoided if a workable indigenous economic model is to be found. All aspects of a nation's economic system (e.g., social, education, and information) must be assessed in the development of the model.

The benefits of having an indigenous economic model are numerous. If successful, it can encourage the country to undertake other self-help projects that might, otherwise, be left to external forces. It can serve as an incentive for citizens to believe in their own economic system. It can force people to exhibit more responsibility in their actions. The nation can exude pride for taking her own destiny in her own hands. The implementation of the model can serve as an example and guideline for other countries in similar economic situations. The economic model can be presented to the citizens as an accurate description of their behaviors. That may compel them to reflect and become more conscious of the way they approach consumption of goods and services, thereby providing avenues for controlling inflation. The model can be an educational tool by making each sector of the economy more appreciative of the rationale behind the actions of other sectors. The idea of an indigenous economic model should provide the directions in which to look for solutions to the pervasive development problems facing the nation. The development of a structure indigenous economic model will be subject to the following categories of factors.

Technology
- Raw materials
- Technology transfer
- Local research
- Skilled labor
- Technology infusion
- Local-context research

Industry diversity
- Oil and gas
- Farming
- Goods and services
- Import–export transactions
- Industrial production facilities

External influences
- Foreign markets
- Technology access
- Trade obligations and restrictions
- Laws and regulations

Public infrastructure
- General public response
- Social climate
- Government agencies
- Political system
- Banking system
- National productivity
- Healthcare system

Education
- Educational system
- Funding sources
- Research engagements
- Study abroad opportunities
- Industry involvement

Products
- Preferences and options
- Quality
- Availability
- Supply and demand
- Cost and inflation
- Brand options

Entrepreneurial opportunities
- Business climate
- Business loan
- Financial resources
- Taxes and incentives
- Business etiquette
- Collusion
- Monopoly

Government
- Bribery and corruption
- Investment climate
- Exchange control
- Immigration system
- External affairs

Conclusion

Product standardization is another factor that can facilitate industrial development. If a standard is available, product interchangeability will be possible. Thus, the market for an industrial product can expand. With an expanded market, better levels of industrialization can be achieved. Units of measure have been one area where standardization has been pursued in many nations. For example, conversion to the metric system has been pursued by several nations as a means to facilitate product compatibility and increase world trade. As more and more countries are switching to metric, a standard system of measurements will simplify international trade. The process of standardization is complicated. But once the initial difficulties of adopting a standard have been overcome, industrial exchanges will become easier. One of the major problems in industrial development is the lack of product integration. We may be effective in making good individual products. But when it comes to fitting the products together to arrive at an overall assembled system, we may have difficulties if there are no prevailing standards. Products should be designed with consideration for how they support one another to achieve an overall workable system in the match toward national industrialization.

References

American Society for Civil Engineers (ASCE), *2013 Report Card for America's Infrastructure*, http://www.infrastructurereportcard.org/ (accessed June 5, 2014), Reston, VA, 2013.

Atta, J. K., *A Macroeconomic Mode of a Developing Economy: Ghana*, University Press of America, Washington, DC, 1981.

Gharajedaghi, J., *A Prologue to National Development Planning*, Greenwood Press, New York, 1986.

National Academy of Engineering (NAE), *14 Grand Challenges for Engineering*, NAE, Washington, DC, 2008.

chapter three

Africa's emerging markets

The awakening of Africa

Africa's economic rise provides evidence that there is a development boom afoot in Africa. In spite of the conventional perception of the continent, Africa is emerging as a viable avenue for global market growth. A large proportion of the growth has manufacturing implications. The focus on other emerging markets, such as India, Brazil, Turkey, Indonesia, Russia, Argentina, and others, has dwindled due to recent sluggish economies of those regions. By contrast, Africa is emerging as more and more vibrant as a business environment.

Africa is the location for five of the world's top 12 fastest-growing economies. The continent has been experiencing comparably strong economic performance from 2000 through 2005. This development is not just about commodity transactions. The benefits of debt relief and improved macroeconomic management have played a role in reversing the economic fortune of Africa. Africa is a large continent in population and it is growing rapidly. This has created a wave of rise in the number of African consumers.

A growing middle class concentrated in urban areas, coupled with a youthful class across the continent, is increasing economic vitality and optimism in Africa. Africa is home to most of the world's least competitive economies. Underdeveloped infrastructure, persistent inequality, political chaos, security, and disease are threatening the recent economic gains. In order to stem the decline and sustain the economic growth, new strategies must be explored. Africa can do both from internal efforts as well as external efforts. Africans in diaspora, who are better educated and more affluent than their predecessors, are wielding tremendous positive impacts on the continent. True to the African culture of communal dedication, as in *it takes a village to raise a child,* Africans in diaspora are not abandoning the continent.

World Bank (2015) data provides the top 12 GDP (gross domestic products) achievers of 2013 in the world as South Sudan (35%), Mongolia (15.3%), Macau (13.5%), Sierra Leone (11.2%), Turkmenistan (9.2%), Bhutan (8.8%), Libya (8.8%), Iraq (8.5%), Laos (8.5%), Timor-Leste (8.5%), Eritrea (8.0%), and Zambia (7.9%). Africa is well represented in the list. A caution needs to be exercised in this list, though. The percentages are good, but represent a growth from a base level. A low base level can experience a huge percentage growth. So, these countries should not rest on their laurels.

More needs to be done in sustaining the growth over the next several years until the base level reaches a point of pride.

GDP is an aggregate measure of production equal to the sum of the gross values of domestic products added up for all local production activities. Taxes and subsidies on products are accounted for in the computational approaches for GDP. GDP estimates are commonly used to measure the economic performance of a whole country or region, but can also measure the relative contribution of an industry sector. GDP is a measure of *value added* rather than sales and revenues. The value-added measure is the value of product outputs minus the value of raw materials that are consumed in producing the products. Because it is based on value added, GDP also increases when an enterprise reduces its use of materials or other resources to produce the same output. So, the industrial lean principle of eliminating waste and focusing on value-added activities adds positively to the creation of GDP. In practice, GDP estimates are used to calculate the growth of the economy from year to year or from quarter to quarter. The pattern of GDP growth is assessed to indicate the success or failure of domestic economic policies and as an index of whether or not an economy is in a recession state. With this understanding of GDP as a value-adding measure, the good rating that Africa is receiving now is, indeed, a big deal. The achievement needs to be celebrated, sustained, and leveraged for more development benefits throughout Africa.

Figure 3.1 shows a graphical representation of the 2013 growth percentages around the world. Average growth numbers are shown for

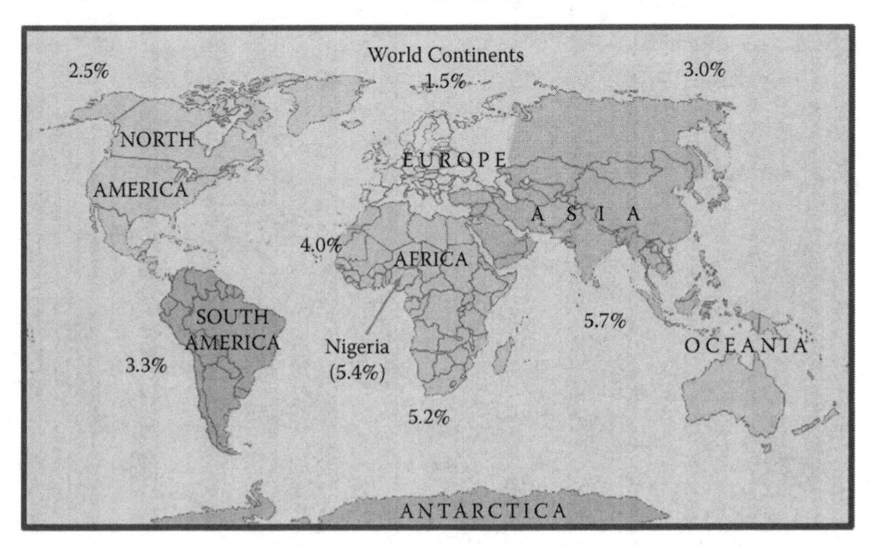

Figure 3.1 The 2013 GDP growth percentages around the world. (Composed and drawn by the author based on World Bank data, http://data.worldbank.org/indicator/NY.GDP.MKTP.KD.ZG.)

selected regions of the world. The African country of Nigeria is specifically called out in Figure 3.1 for a growth rate of 5.4%. Figure 3.2 shows relative 2014 GDPs for selected countries of Africa based on a database from the International Monetary Fund (IMF). With the vibrant GDP growth rates portrayed in Figures 3.1 and 3.2, Africa appears to be the destination for new world markets. Africa, as a continent, is very diverse with spectacular landscape and magnificent people.

Ernst & Young (2013, p.1) asserts in a report that "by 2035, the continent will have the world's largest workforce, with over half of the population currently under the age of 20." The report goes on further to confirm that Africa's rise over the past decade has been very real. While skeptics abound based on pervasive erroneous perception, the evidence of the continent's economic progress over the past decade is well documented. Over the past decade, a critical mass of African economies has grown at a remarkable pace. Despite the prevailing economic downturn globally, the African economy has more than tripled since 2000. The forecast for the future is rosy. The continent, as a whole, is projected to continue experiencing very high growth rates. It is expected that some African countries will remain among the fastest growing in the world for the next several years. This positive trend bodes well for the future of manufacturing in Africa. So, addressing manufacturing technology transfer at this point in time is a timely and prudent undertaking. While pockets of poverty do exist broadly throughout Africa, the core economic indices indicate a promising future for the whole continent. Symbiotic and synergistic influences are needed among the African nations to realize and retain the future potentials. Politics must be set aside for the sake of economic advancement. Adverse perceptions of Africa on the part of the outside world remain a big source of chagrin. It is hoped that more print media representations of the positively increasing economic tide will help change the outside image of Africa.

Pressing Africa's case further, Ernst & Young (2013) goes on to present the five critical success factors for investors in Africa, which are extrapolated as follows:

1. *Perspective*: Assume a glass-half-full perspective that focuses first on the available opportunity. Then, address the attendant risks and how to manage them.
2. *Partnerships*: Invest in building strong collaborative partnerships across government, business, and communities. Supportive communities are the key to a lasting investment success in Africa. They must be brought into the fold using collaborative models similar to the one presented earlier in this book.
3. *Planning*: Adopt careful long-term planning and exercise patience, persistence, and flexibility. These characteristics are essential for business implementations in Africa.

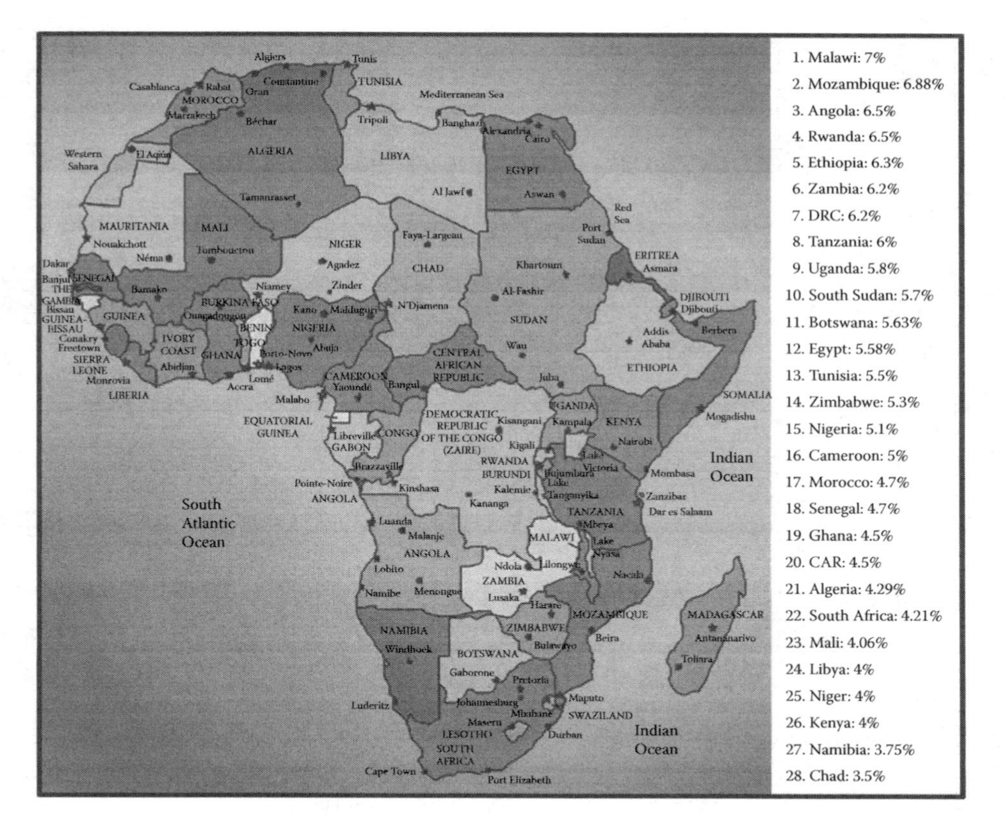

Figure 3.2 The 2014 GDPs across Africa. (Composed and drawn by the author based on IMF data, http://www.imf.org/external/pubs/ft/weo/2014/01/weodata/index.aspx.)

4. *Places*: Embrace Africa's diversity of people, culture, landscape, and other attributes. A systems view of Africa is needed rather than a characterization based on a few case examples.
5. *People*: Participate in celebrating, nurturing, and developing Africa's human resources. The technical and professional success of Africans in diaspora provides a proof of the talents and capability of Africa's human resources.

In a 2014 "Invest in Africa" (2015) campaign, Tullow Oil Plc, in collaboration with Ernst & Young, concluded that "there is compelling evidence that investment is the most effective way to create jobs and expand prosperity. With the mindset increasingly shifting from 'why' to 'how' Africa, we are convinced there has never been a better time to invest; however, unlocking real success and shared prosperity will only come by investors working together with one another, local companies, communities and governments." This is, indeed, very true and it follows the premise of this book.

Ethical principles for global markets

There are issues of diverse cultures and business practices in global market interactions. It is a recognizable fact that many parts of the world face serious ethical challenges with respect to how business transactions are executed. It will be remiss of us to dismiss these issues as inconsequential in global manufacturing technology transfer. To this end, the author recommends being cognizant of the country-to-country issues of bribery and corruption. Nepotism is rampant in many developing countries, where the fear of scarcity (scarcity mentality), issues of being disenfranchised, and concerns about the skewness of the distribution of wealth may lead leaders to act in mendacious ways. It is not uncommon for poor families to expect that own share of the national wealth through crooked practices of a family member who has reached a position of political power and influence at the national level. Authors and practitioners must continue to address these issues squarely, often, and repeatedly. Based on that challenge, the following ethical principles are presented here as a reminder for readers:

1. *Public trust*: Put loyalty to constitution, laws, and ethical principles above private gain.
2. *Conflicting financial interests*: Don't hold financial interests that conflict with performance of duty.
3. *Misuse of nonpublic government information*: Don't use financial transactions to further private interests.
4. *Gifts*: Don't solicit or accept gifts from inglorious sources.
5. *Honesty in efforts*: Put forth the best honest effort in performing your duties.

6. *Unauthorized commitments*: Don't knowingly make unauthorized commitments or make promises purporting to bind your organization to an expectation of reward.
7. *Using public office for private gain*: Don't use your public office for personal private gains.
8. *Impartiality*: Act and make decisions on the basis of merit. Don't exercise preferential or disadvantageous action toward anyone, whether friend, foe, or family.
9. *Custodianship of government property*: Use government property and assets only for authorized and official activities.
10. *Outside employment and compensation*: Do not pursue outside employment and compensation to the detriment of and in conflict with official duties.
11. *Support equal opportunity*: Provide equal opportunity and support for all those deserving it in the performance of your duties.
12. *Appearances*: Avoid actions that create situation in which a reasonable person would question your decision and motives.
13. *Waste, fraud, abuse, corruption*: Avoid waste, fraud, and abuse of government resources and disclose violations to authorities.
14. *Citizen obligation*: Fulfill your obligations as a citizen dedicated to the advancement of the nation. Every little contribution counts.

Recent reemergence of Africa

In a 2010 report at the Center for Global Development, Steven Radelet (Radelet 2010) discussed the 17 countries that are leading the way in the recent reemergence of Africa. There is good news out of Africa, Radelet reports. Seventeen emerging countries are putting behind them the conflict, stagnation, and dictatorships of the past. Since the mid-1990s, these countries have defied the old negative stereotypes of poverty and failure by achieving steady economic growth, deepening democracy, improving governance, and decreasing poverty. The report addressed the political and economic changes under way in the 17 countries. The report also examines three groups of countries:

1. The emerging countries
2. The oil exporters (where progress has been uneven and volatile)
3. The others (where there has been little progress)

There are five fundamental changes under way in the emerging countries, as summarized below:

1. More democratic and accountable governments
2. More sensible economic policies

3. The end of the debt crisis and changing relationships with donors
4. The spread of new technologies
5. The emergence of a new generation of policy makers, activists, and business leaders

Seventeen emerging African countries

According to Radelet (2010), the 17 emerging African countries contain more than 300 million people. Excluding oil exporters, Radelet identifies *emerging countries* as those that achieved an increase in annual income of at least 2% per capita over the period from 1996 to 2008. Zambia barely entered the list. The complete list is presented below:

1. Botswana
2. Burkina Faso
3. Cape Verde
4. Ethiopia
5. Ghana
6. Lesotho
7. Mali
8. Mauritius
9. Mozambique
10. Namibia
11. Rwanda
12. São Tomé and Príncipe
13. Seychelles
14. South Africa
15. Tanzania
16. Uganda
17. Zambia

Six threshold countries

In addition, there are six *threshold countries* that were selected by Radelet because they have shown either promising growth rates over a longer period to time or rapid economic progress in recent years. The six threshold countries are as follows:

1. Benin
2. Malawi
3. Senegal
4. Liberia
5. Kenya
6. Sierra Leone

The emergence of these African countries has been little noticed by the outside world. When noticed, it is often due to negative stereotypical news and reports in the West. The economic turnaround in the 17 emerging countries is evidenced by data: between 1975 and 1995, their economic growth per capita was essentially zero. However, between 1996 and 2008, they achieved growth averaging 3.2% a year per capita, equivalent to overall GDP growth exceeding 5% a year. That growth has powered a full 50% increase in average incomes in just 13 years, according to World Bank reports. Radelet (2010) points to five fundamental changes in the 17 emerging countries, as summarized below:

1. More democratic and accountable governments
2. More sensible economic policies
3. The end of the debt crisis and major changes in relationships with the international community
4. New technologies that are creating new opportunities for business and political accountability
5. A new generation of policy makers, activists, and business leaders

Item 4 in the list above is particularly of interest because of its implication and relevance to the premise of this book.

Africa–USA investment trajectory

The resurgence of business opportunities in Africa has captured the attention of the U.S. investors and the U.S. government in recent years. The 2014 Africa–USA Leaders Summit, which took place in Washington, DC, on August 4–6, 2014, is a case in point. At the summit, President Obama pledged $33 billion in the U.S. private and public assistance to Africa, although he said most of the improvements must come from Africa itself. This bilateral approach is, indeed, welcomed by the theme of this book. Technology transfer must be a two-way street in order for it to be sustainable. "Let's do even more business together," Obama (2014) told delegates to the first USA–Africa Leaders Summit. President Obama told the leaders of about 50 African nations that some of their governments must bolster the rule of law, reform government regulations and root out corruption to promote economic development. Improved security on a continent that has known its share of terrorism is essential, Obama (2014) said, telling delegates, "The future belongs to those who build, not to those who destroy."

In his remarks to the summit's U.S.–Africa Business Forum, President Obama discussed pledges of more than $14 billion by various American businesses for help with projects involving clean energy, aviation, banking, and construction. Coca-Cola will help provide clean water, General

Electric will assist with infrastructure development, and Marriott will build more hotels, the U.S. president said. "The United States is determined to be a partner in Africa's success," he said. "A good partner, an equal partner, and a partner for the long term. We don't look to Africa simply for its natural resources; we recognize Africa for its greatest resource, which is its people and its talents and their potential" (Obama 2014). The president discussed a total of $33 billion in public and private commitments, including $7 billion in new financing to promote U.S. exports and investments in Africa and $12 billion in help from the president's Power Africa initiative involving private-sector partners, the World Bank, and the Government of Sweden.

Further, President Obama spoke of efforts to increase trade between the United States and Africa, expand African electricity and other infrastructure, improve trade between nations within Africa, and help young entrepreneurs and other business leaders get started.

Before his speech, President Obama signed an executive order creating the President's Advisory Council on Doing Business in Africa.

With the promising climate conveyed by the Africa–USA Leaders Summit, it is obvious that Africa must step up to the plate. Visionary proclamations represent one thing while actual sustainable implementations constitute another thing.

As some past efforts have painfully revealed, in confirmation of a Shakespeare quote, "Things sweet to taste prove digestion sour." Utterances that sound palatable at a forum can prove difficult in actual implementation. But it does not need to be that way. The strategies and techniques presented in this book are essential for accomplishing the ideals so well laid out at the summit.

Revitalization of American manufacturing

Manufacturing faces challenges on the global stage. If manufacturing technology transfer is to succeed, each participant at both ends must accelerate to a level of vibrancy. In this regard, there is a current big push to revitalize American manufacturing as released by the U.S. White House in June 2014 (Obama 2014). This bodes well for Africa–U.S. manufacturing technology transfer.

U.S. manufacturing is on the rise, and the U.S. manufacturing sector is as competitive as it has been in decades for new jobs and investment. The manufacturing sector has added 646,000 jobs since February 2010, the fastest pace of job growth since the 1990s. In June 2014, President Obama outlined new actions to accelerate an emerging trend in U.S. manufacturing: new technologies and entrepreneurship in manufacturing that are providing advantages for the United States and helping hardworking Americans get ahead.

A new White House report, "Making in America: U.S. Manufacturing Entrepreneurship and Innovation" (Obama 2014), demonstrates how new game-changing technologies are reducing the cost, increasing the speed, and making it easier for entrepreneurs and manufacturers to translate new ideas into products made in America. These new technologies are already having an impact, with the growth rate in manufacturing entrepreneurship at its fastest pace in over 20 years.

The Obama administration announced new actions by the federal government and new commitments from mayors and local leaders around the country who, following the President's call to action, are investing locally in manufacturing. The president also hosted the first-ever White House Maker Faire, where the president announced new actions by federal agencies and new public–private commitments to spur local entrepreneurship and inspire young people to pursue careers in manufacturing and engineering.

New technologies are lowering the cost and reducing the time required for businesses and entrepreneurs to design, test, and produce new products. Advances in new technologies for rapid prototyping from laser cutters to CNC routers to 3D printers have placed a premium on locating close to American markets and opened new doors to entrepreneurship and innovation in manufacturing. These new technologies can dramatically lower the cost of prototyping in manufacturing, costs that historically have been a barrier to manufacturing startups and to rapid customization at established companies. These emerging technologies in the United States and the renewed focus on manufacturing innovation are already spurring change in manufacturing throughout the United States. Manufacturers have accelerated investment in research and development and manufacturing entrepreneurs are starting new businesses at the fastest rate in over 20 years as reported in various U.S. states, including Ohio, Tennessee, Oklahoma, Texas, New York, and Wisconsin. The author has had the privilege of visiting the DOE's Manufacturing Demonstration Facility at Oak Ridge National Laboratory in Tennessee for collaborative projects in additive manufacturing (3D Printing) and other clean energy technologies.

U.S. manufacturing plays an essential role in supporting and driving American innovation. Manufacturing represents 12% of U.S. GDP, yet accounts for 75% of all U.S. private sector research and development, and the vast majority of all patents issued in the United States.

U.S. manufacturing is more competitive than it has been in decades. Manufacturing output has increased 30% since the end of the recession, growing at roughly twice the pace of the economy overall, the longest period where manufacturing has outpaced U.S. economic output since 1965.

Entrepreneurship in U.S. manufacturing is on the rise, with the rate of growth in manufacturing entrepreneurship at its fastest pace since 1993: The rate of growth in new manufacturing firm openings, a leading indicator of entrepreneurship, has reached its highest levels since 1993. And for the first time since 1999, the number of manufacturing establishments is growing, as new companies form and existing companies expand, resulting in more than 1400 new manufacturing establishments in 2013. American manufacturers have accelerated investments in U.S. innovation. Manufacturers represent 75% of total annual U.S. private sector investment in R&D, having reached an all-time high of $202 billion in 2012, as a result of the acceleration in U.S. R&D intensity from 2007 to now. Established manufacturers, like Ford and GE, are taking advantage of new technologies like rapid prototyping networks to develop new products and increase the rate of innovation within their firms. Defense industries are creating new products to take advantage of the latest technological developments. Similarly, defense research laboratories, particularly the Air Force Research Lab is dedicated to nonclassified technology transfer to spur new manufacturing activities in the civilian sector.

Africa's energy angle

As Albert Einstein said, "Nothing happens until something moves." Energy is what makes everything move and Africa is a major player in world energy profiles. Oil is one of the most powerful avenues of global trade. All facets of our lives are touched by energy on a daily basis in terms of personal uses, commercial operations, industrial activities, and business transactions. Energy is needed to run everything from simple household items, such as cooking stoves, to large industrial complexes, such as iron and steel plants.

After the Middle East, Africa is collectively one of the world's largest oil producers in the world. Depressingly, the continent has not leveraged the power of its oil-producing activities to generate sustainable economic development. But the landscape is changing rapidly. Political turmoil, internal strife, corruption, and ethnic differences have impeded Africa's economic awakening. The potential is enormous. In order to realize the potential, new technological infusion and rigorous management practices must be implemented. This book offers some of the desired tools, techniques, models, and best practices that are essential for turning Africa around. Although there are numerous oil-exporting countries in Africa, it suffices in this book to profile only the top 10 as documented in world statistics as of 2013.

1. *Nigeria*: Nigeria is Africa's biggest oil producer with about 2.2 million barrels being produced a day. This makes the country the number four world exporter of oil. Being fourth in the world is not a trivial position to hold. The country has, however, lagged in capitalizing on the oil revenues. Many oil-producing countries are far ahead of Nigeria in terms of national development.

This has led to widespread disparaging reports about Nigeria's ability to manage and advance itself. It is an issue of wonderment around the world why Nigeria has so much and yet has very little to show for it. Better management of the resources and assets of the country is urgently needed. There is a positive handwriting on the wall. In April 2014, Nigeria was reported to have become the new largest economy in Africa, displacing South Africa on the basis of GDP.

2. *Algeria*: Algeria is the second oil-producing African country. It produces about 2.1 million barrels a day. Algeria had exhibited oil reserves holding of about 12.2 billion barrels of oil in the year 2010. This is a massive amount equaling about 18 years of total world's production of oil. Like many African countries, the oil sector of the country has experienced scandalous incidents that have impeded a smooth operation of the oil-producing activities.
3. *Angola*: Angola is the third top producer of oil in Africa. The country's 1.9 million barrels of production per day places her seventh in the world. Oil and gas revenues anchor Angola's economy. Many foreign investors have ventured into the energy business with the government, thus creating a solid platform that has built the economy to the level of making Angola one of the richest countries in Africa.
4. *Libya*: Libya produces about 1.7 million barrels per day and exports about 1.2 of the total. Libya has seen a lot of change since the demise of Gadaffi. There has been a series of fights between the rebels and the government forces. However, things are rapidly settling down, thus making the country a stable and promising destination for external investors. The political change has affected the oil companies positively and more is expected to be derived from the large investment opportunities.
5. *Egypt*: Egypt is the fifth largest producer of oil in Africa. The country produces 680,000 barrels per day as of 2013. The main spots of production are the Gulf of Suez, the Eastern Desert, the Western Desert, and the Peninsula of Sinai. Egypt has invested in a pipeline that runs from the Gulf of Suez to the city of Sidi Kerir. Oil is one of Egypt's biggest sources of income for the economy.
6. *Sudan*: Sudan is the sixth biggest producer of oil in Africa, producing about 487,000 barrels a day. Sudan has oil as its biggest economic activity. The country has had its share of problems with the North and the South political skirmishes. The eventual splitting of the country into two independent nations initially gave home for a bilateral economic and political advancement, but things are still tense in both areas. However, the oil operations are continuing, thereby creating hope that things will eventually get better.

7. *Equatorial Guinea*: The seventh biggest producer of oil in Africa is Equatorial Guinea, which produces about 346,000 barrels each day. Oil is a major component of the economy in Equatorial Guinea, which enjoys the convenience of having most of its oil rigs located just off the coast.
8. *The Republic of Congo*: The eighth biggest producer of oil in Africa is the Republic of Congo. The country produces about 274,400 barrels of oil per day as of 2013. The country has also been unstable from wars and political upheavals. The stable economy being enjoyed now has been a part of improving the oil sector with more investors considering the country as an investment destination.
9. *Gabon*: Gabon is the ninth biggest producer of oil in Africa. The country produces about 241,700 barrels of oil each and every day. This is a small country (1.71 million population estimate for 2014) with a lot of economic potential due to her oil-producing activities.
10. *South Africa*: South Africa rounds up the list of the top 10 oil producers of Africa. The country produces about 190,000 barrels of oil per day as reported in 2013. South Africa is one of Africa's richest countries. The country is looking into more prospects with the opening up of new oil wells to expand oil and gas businesses. As of 2013, South Africa is a net importer of oil because the country requires about 579,000 barrels of oil per day for internal consumption. The shortfall is imported from other countries.

The list of the top 10 may shift within the next few years because many of other African countries, including Kenya, Uganda, and Ghana, have embarked on aggressive oil exploration activities. Overall, Africa looks very promising in terms of global markets for all sorts of commodities. The international investment climate is getting rosier.

Conclusion

The preceding portrayal of manufacturing and economic vitality in Africa and the United States portend a strong practical tie for Africa–USA manufacturing technology transfer and alliance. Chapters 4 through 15 address specific issues, techniques, tools, and strategies pertinent for manufacturing technology transfer. Intra-Africa trade is impeded by the prevalence of warring relationships and border discords across many African nations. This has the consequence of hindering technology transfer at the continental level. Recent health issues, such as Ebola, have also adversely affected movement of goods, services, and technology across Africa. More collaborative and concerted efforts by African nations are needed to stem the tide of adversities for the sake of putting Africa on a more solid technological footing.

References

Ernst & Young Global Limited, *Africa 2013: Getting Down to Business*, Ernst & Young's attractiveness survey, Times Square, New York, p. 1, 2013.

Invest in Africa, *What Is Invest in Africa?*, 2015. http://investinafrica.com/about-us, (accessed May 13, 2015).

Obama, Barack, *Remarks by the President at the U.S.-Africa Business Forum*, 2014. https://www.whitehouse.gov/the-press-office/2014/08/05/remarks-president-us-africa-business-forum, (accessed May 13, 2015).

Radelet, S., *Emerging Africa: How 17 Countries Are Leading the Way*, CGD Brief, Center for Global Development, Washington, DC, 2010, http://www.cgdev.org/. (accessed December 1, 2014).

Radelet, S., Success Stories from "Emerging Africa," *Journal of Democracy*, Vol. 21, No. 4, pp. 87–101, 2010.

World Bank, *GDP Growth (Annual %)*, World Bank data, 2015. http://data.worldbank.org/indicator/NY.GDP.MKTP.KD.ZG, (accessed May 13, 2015).

chapter four

Global information for industrial development*

Introduction

A chemical society might ask, what on earth is not chemistry? Salisu (2002) asserted for the information sector that "Everything on Earth is Information." Historically, societies and economies had suffered from a lack of information, but information flows at an unprecedented rate so much so that we have thus moved from scarcity to glut. This exponential growth of information output has greatly impacted most human activities in this century. They have become information-driven. The creation, dissemination, and application of information are fundamental to national development in all spheres. While there is an *information glut* in the developed countries, the developing countries are still *information poor.* Nigeria, as an example, belongs to the group of these developing nations.

This chapter looks at the role of global information in various human endeavors for development with particular emphasis on industrial development. It concentrates on the African continent as a case study. Efforts of governments and research institutions at generating information and using global information networks for regional industrial development are reiterated. Some of the constraints to these efforts as well as prospects are discussed.

Globalization

Broadly defined, globalization is a process of rapid economic integration driven by the liberalization of trade, investments, and capital flows, as well as by rapid technological change and the *Information Revolution.* Historically, globalization has been closely associated with the economic transformation of the world, livelihoods, and modes of existence and in politics, a loss in the degree of control exercised locally and in

* Adapted from Salisu, T. M., Global information for industrial development in the third world: The African perspective, *Seminar Presented at the Center for Industrial Development Research (IE-CIDeR),* Department of Industrial and Information Engineering, University of Tennessee, Knoxville, TN, October 9, 2002.

culture, a devaluation of a collectivity's achievements. However, UN Secretary-General Kofi Annan asked on April 3, 2000, "How can we say that the half of the human race which has yet to make or receive a telephone call, let alone use a computer, is taking part in globalization? We cannot without insulting their poverty." Typically, therefore, the end result of globalization to the third world is a process of the redistribution of wealth and opportunity in the wrong direction from the poor to the rich. This is morally indefensible, economically inefficient, and socially unsustainable. What is needed is a system of global governance capable of managing a process of globalization with redistribution in favor of the poor.

In view of the historical origin of globalization, there have always been two sides to the coin. The elite opinion heralds its benefits, while the critics lament the harm it does. If properly managed, however, globalization will ensure the expansion of global linkages, organization of social life on global scale, and growth of global consciousness with the resultant effect of consolidation of *world society*.

The brighter side of globalization should be exploited for the benefit of the entire human race. Global information system is the sunny side of globalization. The rate of development of the information industry has resulted in an emerging global information world. This information has been accepted as a major factor in production, services, empowerment, and a broad array of societal activities. The power of information and its technologies in the developmental process can assist nations that are, hitherto, impoverished to tackle their poverty problem and advance sustainable human development.

Global information

Information is power, a legitimate power, and those who control the generation, processing, storage, and access to information also control the direction the rest of the world goes. The information revolution made possible by the happy marriage of computers and telecommunication circuitry has completely restructured the global socioeconomic equations and led to the transformation of the world into a global, knowledge-based society referred to as the *global village*. The new technologies have led to the development of increasingly knowledge-based systems of trade and production. Borgman (2000) identified the premise and promise of global information infrastructure. The goal is to link the world's telecommunication and computer networks together into a vast constellation capable of carrying digital and analog signals in support of every conceivable information and communication application. He is very optimistic that the infrastructure will do the following (Borgman, 2000):

> Promote an information society that benefits all: peace, friendship, and cooperation through improved interpersonal communications, empowerment through access to information for education, business, and social good; more productive labor through technology-enriched work environments; and stronger economies through open competition in global markets.

The end result of all these is that physical location will cease to matter. More and more human activities will take place online. This will provide access to information for development in all spheres of human endeavor. Labor will become more productive, and economies will be global and more competitive. There is, however, a gap in information infrastructure between the industrialized and nonindustrialized countries. The former is information rich, while the latter is information poor. This gap has to be bridged to ensure international development. The cost of developing and sustaining a global information network is colossal. The focus must therefore be on community rather than individual connectivity.

The third world

This has become synonymous with the countries of the world typified by being technologically less advanced. These are basically in Asia, Africa, and Latin America. They are generally characterized as poor, having economies distorted by their dependence on the export of raw materials to the developed countries in return for finished products. These nations also tend to have high rates of illiteracy, disease, and population growth and unstable governments. Numerically, the Third World dominates the United Nations, but the group is diverse culturally and increasingly economically, and its unity is only hypothetical.

Most rich and developed countries are in the Northern parts of the world, while most poor countries are to the South. The *North* is often used to refer to the industrialized, developed, and rich countries, and the *South* to the poorer, less industrialized ones. However, Australia, although located in the Southern Hemisphere does not typify an underdeveloped country. The economically rising nations, such as Taiwan, South Korea, and Singapore, have very little in common with desperately poor nations of the typical *South*.

Africa is basically a continent of the South. This is so because all the parameters used to measure underdevelopment are present there. Just as Australia may be located in the Southern Hemisphere, but does not belong to the class of poor nations, so also South Africa is located in Africa, but cannot be classified as a developing country. The other parts of Africa south of the Sahara in the class of Third World are usually referred to as Southern Africa. For the purpose of this chapter, South Africa is excluded from this discussion of the poor, nonindustrialized African continent.

The definition so far has been that of some developed nations being expunged from the Third World zones because they do not fit into the description of the countries in these zones. It is, therefore, an irony to have pockets of locations within the very advanced countries that fit into the characteristics of developing nations. The parameters for measuring underdevelopment prevail in these areas. Even within a developed geographical region, pockets of economic discrepancy can exist. A glaring state-side example is the spectrum of economic development that exists from West Tennessee through Middle Tennessee and on to East Tennessee, where industrial development is at its lowest ebb compared to the other end of the spectrum. This trend is being redressed aggressively by the state government. The Industrial Engineering Center for Industrial Development Research (IE-CIDeR) at the University of Tennessee in 2000 through 2006 addressed strategies for such redevelopment efforts.

Industrial development in the third world

It is an incontrovertible fact that industrialization is the bedrock of development in all its ramifications. It is a major index of measuring national development. This is because economic development used to measure the performance of nations and the developmental classification they are given is influenced to a great extent by the level of industrial development. Industry is very important in the development process because it is a source, a user, and a diffuser of technology, which in turn drives sustainable development, productivity, and income growth. Many factors affect industrialization, but a major one is access to relevant, reliable, and timely information. Industrial development, just as any other development process, cannot take place just by opening up (removing tariffs), bringing foreign investment, buying external technology, and enjoying the ride. Efforts have to be made to develop local capabilities in skills, infrastructure, and technology. According to UNIDO (Industrial Development Report 2002/2003. p. 6), "the successful countries are those that have a strategy that combines external drivers and local development." This same report shows that the Third World countries have no place in the committee of nations when it comes to industrial development.

Why underdevelopment

There are so many factors militating against industrial, political, social, and economic development in the African continent. This myriad of problems is mainly human. This is so because the continent is inestimably blessed with natural material resources. The problem is basically that of mismanagement resulting from bad leadership. In actual fact, Africa, particularly Nigeria is also blessed with a high-quality profile of human resources. This is evidenced by the performance of Africans in diaspora who escape from the clutches of the badly managed African continent to thrive professionally on the world stage. The story of African nations can be characterized as that of scarcity in the midst of plenty.

The natural resources in Africa are enormous and enough to make the continent great. But as remarked earlier, the resources are poorly managed to the detriment of the teeming masses. The few people leading the countries use their position to corruptly acquire enormous wealth and impoverish the masses. What we see in Africa, therefore, is growth without development. It is no wonder then that African nations rank high in the list of most corrupt countries of the world. The following factors also militate against an enabling environment for sustainable development.

Poor and unstable polity

Political instability has always been the bane of development in Africa, especially Nigeria. Governance has been very corrupt and unstable. There has been a widespread anarchy, dictatorship, and military gangsterism in the past. The system in operation in these countries and how people survive cannot be comprehended in civilized societies. The political ruling class has always been oppressive and its leadership lacked adequate preparation for governance. It is against this backdrop that development is expected to take place. It is not possible because the external component for a well-rounded development would not get into the country.

Fortunately, the story has started to change for the better in recent years in most parts of the continent, but more remains to be done. Democratic culture is gradually being imbibed, thereby, paving the way for foreign investors into the region, but more yet remains to be done.

Ethnic suspicion and fear

The various ethnic groups in the African continent are suspicious of one another and this has a damning consequence on the rate of development across the board. Various cases of ethnic riots all over the region do not allow for free migration of people and industrial development. A successful industrialist from one part of an African country would not want to establish in some other parts of the country. This is a big risk to take as one is not sure when the next political uprising will develop. Each time there was an ethnic riot, the establishments of the nonindigenes were always in jeopardy. In many cases, loss of lives on a large scale had been recorded.

Security problem

There is a high wave of crime on the African continent with the major cities being most prone to security problems. Life and property are unsafe, and it takes a person of lion heart to invest in activities aimed at developing the region. National security agencies seem not to be able to cope with the security demands. Most industrial outfits commit a great part of their resources to hire private security companies to protect their investments, both human and material.

Economic downturn and debt burden

In the third millennium, the economy of the developing countries remains subsistent. In this era of global economic order and e-business, Nigeria is still grappling with the problems of cash economy. It operates an import–export economy with very little industrialization. There was a great hope seen in taking loans to turn the economy around. Nigeria, like many other developing countries of the world, took loans from Europe, the United States, the World Bank, and the International Monetary Fund. This came with the usual encumbering conditions. This was principally the need for the borrowing countries to make some structural adjustment in their economic policies. This is called structural adjustment programs. Though these programs were originally intended to stimulate growth and provide much needed development aid, they tend to do more harm than good in most of the regions where they have been implemented. This has been accompanied by human damage. What could have been the fortune of this system turned into nightmares for these nations especially the way the funds they took were being grossly mismanaged.

Ferstenfeld (1999) reiterated that "many people assert that the majority of the Third World's debt was brought upon itself by the failure of import substitution and/or the gross mismanagement of funds by corrupt government officials." As much as this may be true, the issue of debt burden has its roots in the larger and inherent unequal international economic system. For example, the Third World has repaid back its debt to the First World many times over in interest payments. And in many cases of course, developing countries have paid more in debt-servicing than the developed countries have given them in aid, loans, and investment. According to *The A–Z of World Development* (Crump and Ellwood 1999),

> From 1983–89 a surplus of $165 billion went from countries receiving aid to the countries that were 'giving' it. Again in 1994, the less-developed countries paid out $112 billion more than they received.

Inadequate infrastructure

The level of infrastructure in most African countries is inadequate and obsolete. The structures required for industrial development are grossly inadequate. These include electricity, water, telecommunications, transportation, and roads.

The supply of electricity is erratic, and when supplied sometimes the voltage is not stable causing damage to domestic appliances and pieces of industrial equipment. A huge portion of the capital of industrial companies is invested in the provision of alternative supply of electricity. This has a negative effect on manufacturing. However, the situation is improving gradually since the advent of a stable polity, but more needs to be done.

The telecommunication sector also suffers the problem of underdevelopment. Many transactions that could have been completed on the phone have to be undertaken physically with a lot of man-hour and resources wasted. For example, the Nigerian government, in the early 2000s, approved private sector participation in the telecommunication industry. Similar strategies have been initiated in some other countries of Africa. The system seems to have worked wonders, in spite of the higher cost to private end users, the consumers.

Transportation and road networks

The road networks in Africa are inadequate to cope with the level of transportation required for any human development. Many manufacturing projects are being thwarted by the lack of adequate facilities for the movement of raw materials and finished products. Huge sums of money are invested to transport goods and services with all the risks involved. Many accidents always result in loss of time and resources causing delay on production lines, in many cases, with the consequent loss of profit.

Information networks

Efforts are being made to ensure that the continent is not left behind in the new information superhighway. Information networks and their interconnectivity are controlling the world, and it is imperative that all the countries of the world must get connected. Computers are now becoming a common place in Africa, most especially in Nigeria. Many private entrepreneurs have entered the business of providing information network services for individuals and institutions. As noble as these efforts are, the result is yet to reach a vast area of the continent. On a positive note, Africa is rated as one of the fastest-growing regions of the world in terms of the use of cellular communication systems. This has facilitated easier and faster banking transactions for the ordinary consumers.

Industrial development institutions

The various countries of Africa recognize the role of industrial development in their societies. They also recognize the role of relevant, reliable, and timely information as well as appropriate research in this development process. They are, therefore, making efforts at generating information through research in their various countries. Also in view of the current wave of globalization, they join international organizations and groups to facilitate development. A good example of such organizations for scientific and industrial development is the World Association of Industrial and Technological Research Organizations (WAITRO). This is an independent, non-for-profit global network of industrial research and technology organizations. WAITRO was established in 1970 to promote and encourage cooperation among industrial and technological research and development organizations. It was founded under the auspices of the United Nations system and has consultative status with many of its specialized agencies. As of 2014, the association has 160 members in 80 countries and cooperates with over 500 other institutes and international agencies worldwide. Over 30 African countries are represented in the membership roster.

As an example, there are six industrial development research institutes in Nigeria, but the following four are the most vibrant:

1. Federal Institute of Industrial Research, Oshodi (FIIRO)
2. Projects Development Institute (PRODA)
3. Leather Research Institute of Nigeria (LERIN)
4. Raw Materials Research and Development Council (RMRDC)

FIIRO

The FIIRO is the oldest research institute set up by the Nigerian Federal Government and was founded in 1956 with the following functions:

- To conduct applied research on Nigerian raw materials to discover their potential industrial uses
- To develop processes to most effectively convert these raw materials into finished products
- To carry out pilot-scale trials of processes found to be technically feasible in the laboratory
- To assess the feasibility of such processes on a commercial scale
- To develop import-substituting products

FIIRO has a wide engineering capability such as design, detailed engineering, fabrication, installation, troubleshooting, and maintenance. It also interacts with many manufacturing subsectors through technical services. It has developed a veritable database called FIIRO INDICES.

PRODA

The PRODA was originally established as an industrial research and development institute by a state government, but taken over by the Federal Government in 1976. Its main aim is to develop industrial projects using local raw materials and indigenous human resources, through laboratory and pilot-plant investigations. PRODA's array of activities includes geological investigations, ceramic research, engineering design, and manufacture of scientific equipment for educational and industrial establishments.

LERIN

The LERIN started off as a project of the UN's Food and Agricultural Organization in 1964. It was, however, taken over by the Federal Government as part of its national science policy. LERIN, among others, is expected to conduct basic and applied research in leather science and technology, investigate vegetable tanning materials, and build up a national information system on leather science and technology.

RMRDC

The *RMRDC* was established as a response to the need to conduct research on local raw materials for manufacturing. The council has a Joint Venture Scheme, which supports individuals or groups of investors with necessary data on specific projects. It has an excellent computerized information system called Raw Material Information System. This council has also turned out a number of relevant publications.

Prospects and summation

It is not as if the developing countries are not aware of what needs be done to keep pace with global development. They are aware and know what to do. They actually make some attempts at institutionalizing policies and practices that will put them on the right track. At the planning stage, very good proposals are designed, but things just go wrong at the implementation stage. This need not be so.

The constraints impeding successful implementation of laudable projects have been highlighted. It is the opinion of the author of this book that the tools and techniques of project management are essential for assuring sustainable implementations of well-crafted development proposals. As such, this book contains a good dose of project management approaches.

Hope seems to be rising in the developing countries of Africa with the efforts being put in to ensure that industrial development, which is crucial to social and economic development, is pursued. The power of global information systems in the development process has been acknowledged by the Third World. As reiterated earlier, many research institutes have been established in Africa to assist in the various activities requisite to industrial development. Computers have become common place in many developing countries. The information systems in these institutes have been connected to the global information system.

Global information systems facilitated by information technologies make possible all sorts of new industrial ventures and new ways of sustaining old activities. The primary goal of global information systems is an equal access to information resources by all who need them. It is anticipated that when this is achieved, people of the whole world will see themselves as partners in progress with little or no suspicion of one another. Vice President Al Gore called for this in his speech on the U.S. Vision for Global Information Infrastructure in 1994:

> Let us build a global community in which the people of neighboring countries view each other not as potential enemies, but as potential partners, as members of the same family in the vast, increasingly inter-connected human family. (Al Gore 1994)

The chagrin is that in spite of similar rhetoric of the past, we are yet to make substantive progress toward a peaceful interconnected global community of the human race. It is envisioned that explicit strategies and techniques presented in this book can begin to move us toward the eventual goal.

References

Annan, K., *Transcript of Press Conference By Secretary-General Kofi Annan at Headquarters, 3 April 2000.* http://www.un.org/press/en/2000/20000403.sgsm7342.doc.html (accessed May 13, 2015), 2000.

Borgman, C.L., *From Gutenberg to the Global Information Infrastructure: Access to Information in the Networked World*, MIT Press, Cambridge, MA, 2000.

Crump, A., W. Ellwood, *The A to Z of World Development*, Oxfam Publishing, Oxford, England, 1999.

Ferstenfeld, M., Structural adjustment: Time for reform/third world countries strangled by debt, *Globalization Newspaper*, Houston, TX, http://cjd.org/1999/02/01/structural-adjustment-time-for-reform-third-world-countries-strangled-by-debt/ (accessed April 7, 2015), 1999.

Gore, A., Speech delivered at the Information Superhighway Summit at UCLA January 11, 1994. http://www.uibk.ac.at/voeb/texte/vor9401.html (accessed May 13, 2015), 1994.

Salisu, T. M., Global information for industrial development in the third world: The African perspective, *Seminar Presented at the Center for Industrial Development Research (IE-CIDeR)*, Department of Industrial and Information Engineering, University of Tennessee, Knoxville, TN, October 9, 2002.

UNIDO, *Industrial Development Report, 2002/2003: Competing through Innovation and Learning*, United Nations Industrial Development Organization, Vienna, Austria, 2003.

chapter five

*Manufacturing or service industry**

There is a misunderstanding between the concepts of manufacturing industry (MI) and service industry (SI). Although the two industrial sectors show similarities and even offer products with common characteristics, these sectors still have important differences. Both developed proper techniques, like supply chain management (SCM) for the MI and yield management (YM) for the SI, and present a tendency to merge their products adapting each other techniques. The goal of this chapter is to contribute to a better understanding of the meaning of MI and SI.

Introduction

The MI is an offspring of the industrial revolution that created the concept of the factory. Along the last century, the MI added technological novelties to its production processes based on scientifically led developments and also added new managerial systems based on scientifically based research studies. The new tendency to form supply chains is a clear spin-off of the new facilities of information technologies and communication that made possible the flow of information and materials along the chain.

The SI is an offspring of commerce. Before the industrial revolution, the craftsman-based production was commercialized very close to the production. In the service, the information technology was essential to define necessary flexibility to allow client-oriented service to customized products and discriminated prices.

The recent facilities provided by automation and information technologies intermingled the activities of MIs and SI. The final costumer acquires products that mix manufacturing aspects—thus more tangible—with the service aspects—thus more personalized.

* Reprinted with permission from A. J. Scavarda and L. J. Lustosa (2004), What are we talking about: Manufacturing or service industry?, *Proceedings of the 20th International Conference on CAD/CAD, Robotics and Factories of the Future*, San Cristobal, Venezuela, pp. 408–417, July 2004.

The techniques that were clearly devised by those two sectors are now rearranged to satisfy the quasimerging of those industries.

The goal of this chapter is to contribute to a better understanding of the meaning of MI and SI. In order to achieve this goal, sections in this chapter present some important concepts of the literature about each industry, discusses how the manufacturing technique, illustrated by the SCM, can be applied on the SI; and how the service technique, illustrated by the YM, can be applied conversely on the MI. The last section of this chapter presents some final considerations.

Manufacturing industry × service industry

Many people believe that the MIs and the SIs are the same thing or that the SI is a specific kind of MI but this is not true! (Thomas 1978; Killeya and Armistead 1983; Armistead et al. 1986; Schmenner 1986; Morris and Johnston 1987; Gummesson 1994; Sampson 2001).

As the SI is people intensive, this industry is more exposed to human mistakes that risk the consumer expectation (Levitt 1981). It is also not trivial to correct the error before the service is consumed, since in this industry, the *product production* stage happens almost simultaneity with the consumption one. In MI, goods can usually be produced well in advance to consumption. These issues add difficulty to set standards and measure quality, and also, they usually make the correction of errors a much more critical task than in manufacture.

In the MI, it is possible to check and control the products before their arrived to the final consumer: in a bicycle industry, for example, it is possible to check if the bikes have two wheels; in the car industry, it is possible to check out if the seatbelts are operating; the beverage industries have techniques that permit them to fill the bottles to a specified volume. Absolute precision does not exist, but in the MI it is possible to meet the specification even if small mistakes are committed.

The SI can control its production procedures, and so its quality, by selecting and training its workers and the procedures before their execution, but the SI will be always jeopardized closer to mistakes, many from customers but also others from unforeseeable reasons. Quality is a very important service issue, but it is difficult to establish and control it on the SI.

The SI deals with intangible values, that means: it deals with values that cannot be easily seen, touched, smelled, heard, or even tasted before the purchase. Many aspects influence the services values, some are defined by people curiosities or wishes of knowledge, some are influenced by specific groups and others are influenced by emotional feelings. Besides this, some values are based on conditional parameters, where temporary situations enhance different reasons to consume.

The SI is intensively commanded by the consumer, thus it is very important to understand the consumer expectations. This understanding underlines the ability to perceive consumer's feelings (if the consumer is happy or disappointed), his attractions to the service (if he is going to buy or not the service), and maybe the consumer mood (when the customer is in a bad humor, he tends to be less tolerant).

Consumers have many different reasons to buy or use a service. It is not easy to control the service demand, since its structure involves many parameters: internal, external, and firm-produced factors. The internal factor involves personal needs and asks for past experiences and the involvement of the consumer, while external factor involves competitive options and the social contexts. The firm factor involves the firm attitudes: promotions, pricing, distributions, and alliances. It is important to emphasize that people and the environment (attractive decor, appealing smells, music and friendly service personnel) have a decisive influence.

The SI also involves many other delicate points, like the quality evaluation. Quality is very subjective in an intangible product. Tough, in the environment of services, the procedures, the way the personnel acts and many other aspects of the service can be poorly controlled; moreover, the final quality criterion will always be the consumer perception. The consumer perceived quality is not accurately predictable and it is particular to each consumer. Sometimes the responsiveness is more important to the client than simply the offer of the *best* service; the client wants to feel that the people involved are trying to do their best to serve him or her. The basic difference between MI and SI in terms of quality comes from the difficulty to define quality and then to parameterize quality under the given definition.

The quality is very important, but it is mostly noticed when it is not present. Some SIs try to show the quality of the product (like sending letters to the clients with information and felicitations), in general, they try to make tangible what is really intangible. In reality, all products have tangible and intangible values and each value can and should complement the other.

Firms and their employees have a fundamental role in many issues that impacts quality. Some research studies are being done on this subject; Hays and Hill (2001) developed a longitudinal model to study the ascendance of the service quality, mainly on the firm's service guarantee point of view. This study provided insights into the mechanism by the service guarantee effects on the service quality. It analyzed two paths strongly connected with the employee actuation: employee motivation and vision, and learning through service failure.

However, in the literature, many research studies done over the MI, like SCM, could be applied over the SI, but for that it is necessary to base them on the SI requirements and not on the manufacturing ones.

Analogically the same could be mentioned about the service influence over the SCM, such as the YM.

SCM: An example of a manufacturing technique that can be applied on the SI

There are many definitions to SCM, most of them related with the MI. According to Pires (1998), SCM is a contemporary strategic approach that manages manufacturing as a whole. The basic objective of this management is to maximize the synergy between all segments of the demand chain in order to serve the end consumer more effectively, by either reducing costs or enhancing value. In order to achieve this, companies have sought to align and join distinctive competencies along the demand chain. Competition is now focused on productive chains instead of on single companies and this can provide many of the benefits of vertical integration without losses in cost and flexibility (Vollmann and Cordon 1996).

Among the SCM conceptual frameworks proposed in the literature, the one by Lambert and Cooper (2000) merits special attention here because it addresses implementation issues and is based on empirical evidences. This framework is composed by three basic elements: the network structure, business processes (customer relationship management, customer service management, demand management, customer order fulfillment, manufacturing flow management, procurement, product development and commercialization and returns), and management component (planning and control of operations, work structure, organizational structure, product flow facility structure, information flow facility structure, management methods, power and leadership structure, risks and rewards, and culture and attitudes). Each element corresponds to a question concerning the analysis and synthesis of SCM, which are as follows: (1) the network structure—who are the key SCM members with whom to link processes?; (2) the business processes—what processes should link each of these SCM members?; (3) the management components—what level of integration and management should be applied for each process link? As the Pires' definition, this framework was also based on the manufacture paradigm.

Though no definition of SCM is universally accepted, its characterization usually involves the idea of formally managing, across different companies, from raw material suppliers to final consumers, the entire chain of activities necessary to produce value to the final consumer. Of course, this integrative management has the objective of making the chain's product more competitive, in other words, of lower cost, higher quality, more dependable, more diversified, and promptly delivered.

Extending the SCM to the SI

Basically, in both type of industry (manufacturing and service), the SCM is a process that guaranties simultaneously the mass production and customization. From the Fordism period to the present, the production has changed becoming more customer driven. Nowadays, to be competitive many industries need to have mass production based on flexible systems and Internet ordering in order to compete in price, quality, product flexibility, and fast and dependable delivery (Solihull 2001). Mass customization is key to the SI, where, with a fast and flexible system, it is always possible to change things to rise together both customer satisfaction and producer profits. In service, flexibility is largely related to employee empowerment. The key for this flexibility is the employee motivation and vision and the organizational learning that may influence the service quality that will stimulate customer satisfaction, and consequently business performance and profit (Hays and Hill 1999).

In spite of the importance of SCM to the MI and SI, it is still necessary to effectively extend its analyses to the SI. A vast literature on SCM has been produced in the last decade, but excluding sale services and other few exceptions (Fitzsimmons 2000) most of this literature addresses the manufacturing case with little attention to SI chain.

This lack of attention is somewhat surprising. First of all, in the developed economies, the service sector is expanding fast and has become the largest one. Second, production and consumption are simultaneous in services, making co-ordination of these activities among the suppliers more critical than in manufacture. At last, the progress in the so-called Information Technologies (ITs), including e-commerce, and integrated enterprise management systems, that allows and promotes SCM integration, affect essentially the managerial part of the organizations in such a way that services should profit from them no less than manufactured products.

It is important to be able to visualize the SI chain structure with the reference of the MI. While the MI chain is characterized by its strong sequential organization, the SI chain has a net format. Very often the service chain becomes a service net. This means that in the service chain the final customer can have contact with many members of the chain, which is unusual in the manufacturing chain where the final customer generally does not have contact with the product's distributor or assembler.

As the SI contains many peculiar characteristics that are very different from the MI, many aspects here should be considered and analyzed in order to extend the supply chain focus to a service chain or net focus. In the SI chain, for instance, the distinction between product and production process is blurred due to its product intangibility. Many factors that influence the performance of the SI and stimulate the service chain integration should be studied.

The supply chain structure involves a combination of many different activities and products, including many different companies and members of the chain. Probably the idea of supply chain arise from the coalition between some business creating the condominium (joint ownership), then the mall, and finally the integration of the industries. The basic supply chain structure is composed of a member responsible for the integration of the whole chain—the chain coordinator (that *assembles* the final product), the final customer, and the rest of the chain. The structure of the industry chain involved (including its processes, functions, management components, and links) and the levels of interactions and partnership between its members will result on a synergy of its activities. An analogy can be done between the MI and the SI. It is possible to extend the perception of the manufacturing supply chain structure, here represented by the automotive segment, to a service supply chain structure, represented by the health care, education, the bank, and the tourism industry.

In the manufacturing case, the automotive segment, the vehicle assembler coordinates the participation of many other MIs in order to build a car with low costs and high quality. Some differences of assembling the car are done, making it possible to produce a family of models, allowing the car buyer to choose a variety of models, but not so many.

In the service case, there are many likenesses with the manufacturing case. In the hospital field, each patient is a different case, many times the patient does not know what he or she needs and cannot immediately be exposed to the final and complete product. The general practitioner is the one responsible to assemble the product. He or she coordinates the laboratorial exams, medicines, the surgical processes, the hospital internment, and the recovery process. In the school/university field, each student has personal interests and learning speed. The teacher (at school) and the adviser (at university) help the student to connect his or her desires with the huge quantity of disciplines, apprenticeship, research programs, and laboratories. In the banking segment, each investor wants to apply his or her money in a different way. Some people are aggressive (enjoy risks) and others are prudent, preferring stable investment. The bank manager helps them to compose their portfolio, providing the clients with actual information about each company. The chain is not just composed by other companies but also by controller and regulator institutes. Also the bank manager may assist the current account holder, or any other account holder, to find the best ways to execute their specific payments. In the tourism case, each person has different dreams and wants to visit different places and has different kind of experiences. The tourism operator and the tourism agency compose a different tourism package to the tourist that can include restaurants, hotels, airplanes, and amusements, among others.

YM: An example of a service technique that can be applied on the MI

YM is composed basically of pricing and inventory control. It deals with an uncertain demand for a service whose capacity is rather fixed and has a high fixed cost. Airplane seats in a flight, hotel rooms available at a given day, and seats in a theater performance are just a few examples of this situation. YM is a method to maximize the revenues of an industry with limited capacity that considers the forecast of demand and that tries to spread the demand along the year and along the physical capacity of the industry. As it is able to treat groups individually, it can also maximize the revenue by segmenting the market in groups of clients, charging particular prices, and offering services to groups of people that were excluded by the market (Lovelock 1992).

The segmentation of the market is important to the industry and also to the consumer. Each person has different habits, necessities, and tastes, so each client requires a different product. There is a specific technique, the customer profitability analysis (CPA), that examines the revenues and cost for each different customer or group (Noone and Griffin 1999). To be able to personalize the service means to be able to defend the customer individuality guaranteeing a democratic market. The segmentation can allow the revenue maximization. Each customer is responsible for different revenue to the industry. No customer is identical to another, even when requiring a similar product at the same time. The customers will be willing to pay more or less for a service according to their necessities and the way they want that service to be executed.

YM is a class of methodologies that helps to cope with the above problem. Its intensive use started by the airline companies and, now, reached also the hotel and the rental car industries. This method is now being used also by the rest of the SI when limited capacity defines difficulties in maximizing the real market possibilities; examples are the entertainment industry (e.g., stadium, park, theater, and cinema), commercial centers, hospitals, schools, restaurants, night clubs, the transport industry, and the tourism industries.

In spite of its great success around the world, mainly in the airline industry, some industries consider the YM just an American theory, in other words: they think that this technique is only good inside the United States. This is the case of the hotel industry, mainly in the United Kingdom, that presented difficulties in using the YM. One of the problems is to train the staff to use YM correctly. The staff is not just worried to have to learn a new technique, but mostly is confident on the method because [the staff] believes that it represents an extra load to the job.

Without the YM the staff was allowed to ask any price for a room. With the YM system, the negotiation started to be more realistic and the YM define what the trade-offs should be.

YM should not substitute completely the human guess, the feeling of the worker is also very important and should be considered as a fundamental parameter among the YM inputs. The workers should understand better how does the YM work, how it could affect their work, and how they can interfere on it. The YM should consider the peculiarities of the industry or industries on which it is being applied.

In the air SI, it needs some special attention with parameters such as multiple legs, booking levels prior to departure, boarding by fare class, choice behavior (in spite of the demand for each fare class is distinct, the passenger can choose another fare class or change his flight), and no-shows. No-shows consist on the fact that not always the reservation is converted in its respective use of the service. This is the real case of the overbooking system. It is not simple to implement overbooking because the industry has to minimize the risk of denying passengers to travel.

The inventory control can be done by separate and nested fare class inventories. It is necessary to do a complete inventory of the seats in order to reach its optimum number. Depending on the problem complexity, it might be necessary to use differential calculus, Lagrangian multipliers, mathematical programming, and network.

It is also important to decide if the problem is static or dynamic. In the static problem, at the beginning of the booking, fare class booking limits are applied for the next flight considering the uncertainty associated with expected bookings by fare class. In the dynamic problem, while booking is accepted, the booking limits are recalculated revising the initial booking limits (Belobaba 1989).

Many of the air industry problems, like no-show, also exist in the hotel industry, and vice-versa. In the hotel case, there are some slight differences in the necessities of the application of YM. The hotels, in comparison with the air companies, have several competitors. The guests, who do not initially accept to pay more than another guest for *the same room* can just cross the street and choose other hotels. So, the worker must explain the guests that the other guest, the one who had paid, was part of a previous corporate agreement or had been booked at a *special offer* (Bradley and Ingold 1993). The problem does not comes from complains, but from customers that do not accept the different prices and do not complain. These ones will be disappointed, will probably not return to that hotel again, and will dissuade others to do so. To apply the YM, then, the hotel industry does not only face the problem to train its workers but also to educate the consumer for this kind of management that can initially sound strange. There are other problems like the multiplier functions of rooms on other hotel functions (such as restaurant, stores, souvenirs, tours and convention space), the booking lead times for various types of rooms, the lack of a distinct rate structure, and decentralized information systems (Kimes 1989).

Extending the YM to the MI

Many aspects of the YM can be useful for the MI (Weatherford and Bodily 1992; Shin and Park 2000; Elimam and Dodin 2001; Belobaba 2002; Jarvis 2002; Hawtin 2003). YM involves segmenting customers, setting prices, and controlling capacity to maximize the revenue produced with a fixed capacity.

The MIs have their production capacity limited according to their equipments and working hours. Many MIs are already using or renting their equipments according to the importance and value of their products.

Nowadays, with the Internet, many MIs are using pricing mechanisms, such as auctions, guaranteed purchase contracts, and group purchasing (Vulcano et al. 2002). The prices of the manufacturing products are declining very fast, and so the goal is to introduce the product to the market as soon as possible, mainly in a technology-rich industry (Lee 2000). In the electronic marketplaces the buyers use dynamically arriving e-orders and by the Internet they post the prices that they want for a specific product. The MIs, the price takers, decide if they will accept the order at the time of its arrival, and, if so, they must define how they will manage the production, control the inventory levels, and balance the benefits and costs associated with providing the good to the end customers by controlling the admission of orders and the levels of stock (Caldentey and Wein 2002).

Final considerations

The MI was initially developed along the twentieth century with vertical enterprises able to deliver products manufactured since the choice of raw materials up to the final sale. This is fundamentally a production-pushed process led by the industry.

The last two decades of the twentieth century changed the vertical company in a more flexible set of smaller companies, changing the production into a demand-pulled supply chain led by the customer. This was the basic mechanism that bought service aspects to a basically MI. On the other hand, the fast technological development making manufacturing goods very cheap and of high quality reshaped the already customer-led services, making practically impossible to deliver services free of manufacturing goods. This was the basic mechanism that brought manufacturing aspects to a basically SI.

This chapter does not intend to make a thorough study about MI and SI, but intends to contribute to this cross-study by discussing how SCM-manufacturing technique can be applied on the SI and how the YM service technique can be applied on the MI. Many facts about the cross-studies of these areas are based on anecdotal beliefs assumed to be correct, and a thorough research effort to comprehend these issues is still missing.

References

Armistead, C., R. Johnston, C. A. Voss, Introducing service industries in operations management teaching, *International Journal of Operations and Production Management*, Vol. 6, No. 3, pp. 21–29, 1986.

Belobaba, P. P., Application of a probabilistic decision model to airline seat inventory control, *Operations Research Society of America*, Vol. 37, No. 2, pp. 183–197, 1989.

Belobaba, P. P., Future of revenue management, *Journal of Revenue and Pricing Management*, Vol. 1, No. 1, pp. 87–89, 2002.

Bradley, A., A. Ingold, An investigation of yield management in Birmingham Hotels, *International Journal of Contemporary Hospitality Management*, Vol. 5, No. 2, pp. 13–16, 1993.

Caldentey, R. A., L. M. Wein, Revenue management of a make-to-stock, *Queue Manufacturing and Service Operations Management*, Vol. 4, No. 1, pp. 4–6, 2002.

Elimam, A. A., B. M. Dodin, Incentives and yield management in improving productivity of manufacturing facilities, *IIE Transactions*, Vol. 33, No. 6, pp. 449–462, 2001.

Fitzsimmons, J. A., Service chain management, *Proceedings of the First World Conference on Productions and Operations Management POM Sevilla 2000*, Spain, August 26–30, 2000.

Gummesson, E., Service management: An evaluation and the future, *International Journal of Service Industry Management*, Vol. 5, No. 1, pp. 77–96, 1994.

Hawtin, M., The practicalities and benefits of applying revenue management to grocery retailing, and the need for effective business rule management, *Journal of Revenue and Pricing Management*, Vol. 2, No. 1, pp. 61–67, 2003.

Hays, J. M., A. V. Hill, The market share impact of service failures, *Production and Operations Management*, Vol. 8, No. 3, pp. 208–220, Fall 1999.

Hays, J. M., A. V. Hill, A longitudinal study of the effect of a service guarantee on service quality, *Production and Operations Management*, Vol. 10, No. 4, pp. 405–423, 2001.

Jarvis, P., Introducing yield management into a new industry, *Journal of Revenue and Pricing Management*, Vol. 1, No. 1, pp. 67–75, 2002.

Killeya, J. C., C. G. Armistead, The transfer of concepts and techniques between manufacturing and service systems, *International Journal of Operations and Production Management*, Vol. 3, No. 3, pp. 22–28, 1983.

Kimes, S. E., The basics of yield management, *Hotel Operations—The Cornell H.R.A. Quarterly*, Vol. 4, pp. 14–19, 1989.

Lambert, D. M., M. C. Cooper, Issues in supply chain management, *Industrial Marketing Management*, Vol. 29, pp. 65–83, 2000.

Lee, F., Yield management: Present and future, *Semiconductor International*, Vol. 23, No. 3, pp. 85–90, 2000.

Levitt, T., Marketing intangible products and product intangibles, *Harvard Business Review*, pp. 94–102, May–June 1981.

Lovelock, C. H., Yield management: A tool for capacity-constrained service firms, *Managing Service—Marketing, Operations, and Human Resources*, Prentice-Hall International, Englewood Cliffs, NJ, pp. 188–201, 1992.

Morris, B., R. Johnston, Dealing with inherent variability: The difference between manufacturing and service? *International Journal of Operations and Production Management*, Vol. 7, No. 4, pp. 13–22, 1987.

Noone, B., P. Griffin, Managing the long-term profit yield from market segments in a hotel environment: A case study on the implementation of customer profitability analysis, *International Journal of Hospitality Management*, Vol. 18, pp. 111–128, 1999.

Pires, S. R. I., Managerial implications of the modular consortium model in a Brazilian automotive plant, *International Journal of Operations and Production Management*, Vol. 18, No. 3, pp. 221–232, 1998.

Sampson, S. E., The unified services theory approach to service operations management, *POM 2001 Orlando—POM Mastery in the New Millennium*, Florida International University, Orlando, FL, March 30–April 2, 2001.

Schmenner, R. W., How can service businesses survive and prosper? *Sloan Management Review*, Vol. 27, No. 3, pp. 21–32, Spring 1986.

Shin, C. K., S. C. Park, A machine learning approach to yield management in semiconductor manufacturing, *International Journal of Production Research*, Vol. 38, No. 17, pp. 4261–4270, 2000.

Solihull, N., Special report mass customization—A long march, *The Economist*, July 14, 2001, pp. 67–69.

Thomas, D. R. E., Strategy is different in service businesses, *Harvard Business Review*, Vol. 56, No. 4, pp. 158–165, 1978.

Vollmann, T., C. Cordon, Making supply chain relationship work, *M2000 Business Briefing, No. 8*, IMD, Lausanne, Switzerland, pp. 4–5, 1996.

Vulcano, G., G. van Ryzin, C. Maglaras, Optimal dynamic auctions for revenue management, *Manufacturing and Service Operations Management*, Vol. 4, No. 1, pp. 7–11, 2002.

Weatherford, L. R., S. E. Bodily, A taxonomy and research overview of perishable-asset revenue management: Yield management, overbooking, and pricing, *Operations Research*, Vol. 40, No. 5, pp. 831–844, 1992.

chapter six

Manufacturing processes and systems*

Definition of manufacturing and its impact on nations

In his book *Manufacturing Systems Engineering*, Hitomi (1996) differentiated between the terms *production* and *manufacturing*. According to him, production encompasses both making tangible products and providing intangible services while manufacturing is the transformation of raw materials into tangible products. Manufacturing is driven by a series of energy applications, each of which causes well-defined changes in the physical and chemical characteristics of the materials (Dano 1966).

Manufacturing has a history of several thousand years and may impact humans and their nations in the following ways (Hitomi 1994):

- *Providing basic means for human existence*: Without manufacture of products and goods humans are unable to live, and this is becoming more and more critical in our modern society.
- *Creating wealth of nations*: The wealth of a nation is impacted greatly by manufacturing. A country with a diminished manufacturing sector becomes poor and cannot provide a desired high standard of living to its people.
- *Moving toward human happiness and stronger world's peace*: Prosperous countries can provide better welfare and happiness to their people in addition to stronger security while posing a less of a threat to their neighbors and each other.

In 1991, the National Academy of Engineering/Sciences in Washington, DC, rated manufacturing as one of the three critical areas necessary for America's economic growth and national security, the others being science and technology (Hitomi 1996). In recent history, nations which became active in lower level manufacturing activities have grown into

* Reprinted with permission from Badiru, Adedeji B. (2014), Handbook of Industrial and Systems Engineering, Second Edition, Chapter 21 (A. Sirinterlikci), pp. 371–397, CRC Press/Taylor & Francis, Boca Raton, FL.

higher level advanced manufacturing and stronger research standing in the world (Gallager 2012).

As the raw materials are converted into tangible products by manufacturing activities, the original value (monetary worth) of the raw materials is increased (Kalpakjian and Schmid 2006). Thus, a wire coat hanger has a greater value than its raw material, the wire. Manufacturing activities may produce *discrete products* such as engine components, fasteners, gears, or *continuous products* like sheet metal, plastic tubing, conductors which are later used in making of discrete products. Manufacturing occurs in a complex environment that connects multiple other activities such as product design, process planning and tool engineering, materials engineering, purchasing and receiving, production control, marketing and sales, shipping, and customer and support services (Kalpakjian and Schmid 2006).

Manufacturing processes and process planning

Manufacturing processes

Today's manufacturing processes are extensive and continuously expanding while presenting multiple choices for manufacturing a single part of a given material (Kalpakjian and Schmid 2006). The processes can be classified as traditional and nontraditional before they can be divided into their mostly physics-based categories. While much of the traditional processes have been around for a long time, some of the nontraditional processes may have been in existence for some time as well, such as in the case of electro-discharge machining (EDM), but not utilized as a controlled manufacturing method until a few decades ago.

Traditional manufacturing processes can be categorized as

1. Casting and molding processes
2. Bulk and sheet forming processes
3. Polymer processing
4. Machining processes
5. Joining processes
6. Finishing processes

Nontraditional processes include

1. Electrically based machining
2. Laser machining
3. Ultrasonic welding
4. Water-jet cutting
5. Powder metallurgy
6. Small-scale manufacturing

7. Additive manufacturing
8. Bio-manufacturing

Process planning and design

Selection of a manufacturing process or a sequence of processes depends on a variety of factors including the desired shape of a part and its material properties for performance expectations (Kalpakjian and Schmid 2006). Mechanical properties such as strength, toughness, ductility, hardness, elasticity, fatigue, and creep; physical properties such as density, specific heat, thermal expansion and conductivity, melting point, and magnetic and electrical properties; and chemical properties such as oxidation, corrosion, general degradation, toxicity, and flammability may play a major role in the duration of the service life of a part and recyclability. The manufacturing properties of materials are also critical since they determine whether the material can be cast, deformed, machined, or heat treated into the desired shape. For example, brittle and hard materials cannot be deformed without failure or high energy requirements; whereas they cannot be machined unless a nontraditional method such EDM is employed. Table 6.1 depicts general manufacturing characteristics of various alloys and can be utilized in the selection of processes based on the material requirements of parts.

Each manufacturing process has its characteristics, advantages, and constraints including production rates and costs. For example, conventional blanking and piercing process used in making sheet-metal parts can be replaced by its laser-based counterparts if the production rates and costs can justify such a switch. Eliminating the need for tooling will also be a plus as long as the surfaces delivered by the laser cutting process is comparable or better than that of the conventional method (Kalpakjian and Schmid 2006). Quality is a subjective metric in general (Raman and Wadke 2006). However in manufacturing, it often implies *surface finish*

Table 6.1 Amenability of Alloys for Manufacturing Processes

	Amenability for		
Type of alloy	Casting	Welding	Machining
Aluminum	Very high	Medium	High to very high
Copper	Medium to high	Medium	Medium to high
Gray cast iron	Very high	Low	High
White cast iron	High	Very low	Very low
Nickel	Medium	Medium	Medium
Steels	Medium	Very high	Medium
Zinc	Very high	Low	Very high

and tolerances, both dimensional and geometric. The economics of any process is again very important and can be conveniently decomposed with the analysis of manufacturing operations and their tasks. A manufactured part can be broken into its features, the features can be meshed with certain operations, and operations can be separated into their tasks. Since several possible operations may be available and multiple sequences of operations coexist, several viable process plans can be made (Raman and Wadke 2006).

Process routes are sequence of operations through which raw materials are converted into parts and products. They must be determined after completion of production planning and product design according to the conventional wisdom (Hitomi 1996). However, newer concepts like Concurrent or Simultaneous Engineering or Design for Manufacture and Assembly (DFMA) are encouraging simultaneous execution of part and process design and planning processes and additional manufacturing-related activities. Process planning includes the two basic steps (Timms and Pohlen 1970):

1. *Process design* is a macroscopic decision-making for an overall process route for the manufacturing activity.
2. *Operation/task design* is a microscopic decision-making for individual operations and their detail tasks within the process route.

The main problems in process and operation design are analysis of the workflow (flow-line analysis) for the manufacturing activity, and selecting the workstations for each operation within the workflow (Hitomi 1996). These two problems are interrelated and must be resolved at the same time. If the problem to be solved is for an existing plant, the decision is made within the capabilities of that plant. On the contrary, an optimum workflow is determined, and then the individual workstations are developed for a new plant within the financial and physical constraints of the manufacturing enterprise (Hitomi 1996).

Workflow is a sequence of operations for the manufacturing activity. It is determined by manufacturing technologies, and forms the basis for operation design and layout planning. Before an analysis of workflow is completed, certain factors have to be defined including precedence relationships and workflow patterns. There are two possible relationships between any two operations of the workflow (Hitomi 1996), which are as follows:

1. A partial order, *precedence,* exists between two operations such as in the case of counterboring. Counterboring must be conducted after drilling.
2. No precedence exists between two operations if they can be performed in parallel or concurrently. Two set of holes with different sizes in a part can be made in any sequence or concurrently.

Harrington (1973) identifies three different workflow patterns: *sequential* (tandem) process pattern of gear manufacturing, *disjunctive* (decomposing) pattern of coal or oil refinery processes, and *combinative* (synthesizing) process pattern in assembly processes.

According to Hitomi (1996), there are several alternatives for workflow analysis depending on production quantity (demand volume, economic lot size), existing production capacity (available technologies, degree of automation), product quality (surface finish, dimensional accuracy, and tolerances), and raw materials (material properties, manufacturability). The best workflow is selected by evaluating each alternative based on a criterion that minimizes *the total production (throughput) time*, or *total production cost*, defined in (6.1) and (6.2). *Operation process or flow process charts* can be used to define and present information for the workflow of the manufacturing activity. Once an optimum workflow is determined, the detail design process of each operation and its tasks are conducted. A *break-even analysis* may be needed to select the right equipment for the work-station. Additional tools such as *man–machine analysis* as well as *human factors analysis* are also used to define the details of each operation. *Operation sheets* are another type of tool used to communicate about the requirements of each task making up individual operations.

$$\text{Total production time} = \sum \begin{bmatrix} \text{transfer time between stages} \\ + \text{ waiting time} + \text{set-up time} \\ + \text{ operation time} + \text{inspection time} \end{bmatrix} \tag{6.1}$$

$$\text{Total production cost} = \text{material cost} + \sum \begin{bmatrix} \text{cost of transfer between stages} \\ + \text{ set-up cost} + \text{operation cost} \\ + \text{ tooling cost} + \text{inspection cost} \\ + \text{ work-in-process inventory cost} \end{bmatrix} \tag{6.2}$$

where:

$\sum$ represent all stages of the manufacturing activity

Industrial engineering and operations management tools have been used to determine optimum paths for the workflow. Considering the amount of effort involved for the complex structure of today's manufacturing activities, computer-aided process planning (CAPP) systems have become very attractive in order to generate feasible sequences and to minimize the lead time and nonvalue-added costs (Raman and Wadke 2006).

Traditional manufacturing processes

Casting and molding processes

Casting and molding processes can be classified into the following four categories (Kalpakjian and Schmid 2006):

1. *Permanent mold based*: Permanent molding, (high and low pressure) die-casting, centrifugal casting, and squeeze casting
2. *Expandable mold and permanent pattern based*: Sand casting, shell mold casting, and ceramic mold casting
3. *Expandable mold and expandable pattern*: Investment casting, lost foam casting, and single-crystal investment casting
4. *Other processes*: Melt-spinning

These processes can be further classified based on their molds: permanent and expandable mold-type processes (Raman and Wadke 2006). The basic concept behind these processes is to superheat a metal or metal alloy beyond its melting point or range, then pour or inject it into a die or mold, and allow it to solidify and cool within the tooling. Upon solidification and subsequent cooling, the part is removed from the tooling and finished accordingly. The expandable mold processes destroy the mold during the removal of the part or parts such as in sand casting (Figure 6.1) and investment casting (Figure 6.2). Investment casting results in better surface finishes and tighter tolerances than sand casting. Die-casting (Figure 6.3) and centrifugal casting processes also result in good finishes, but are permanent mold processes. In these processes, the preservation of

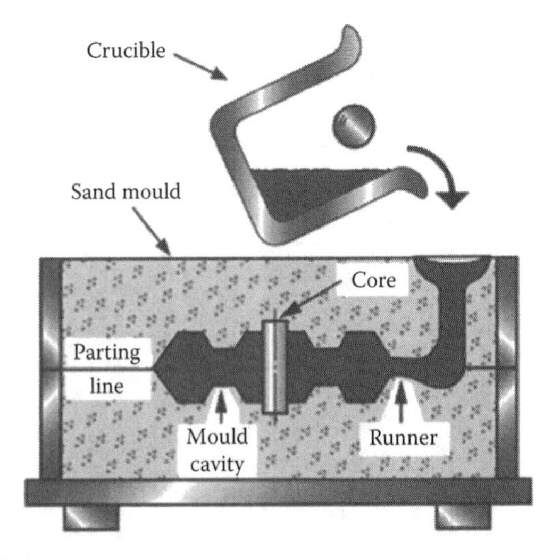

Figure 6.1 Sand casting.

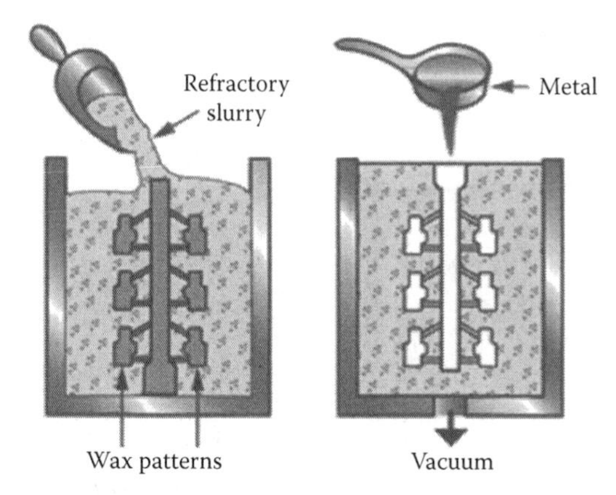

Figure 6.2 Investment casting.

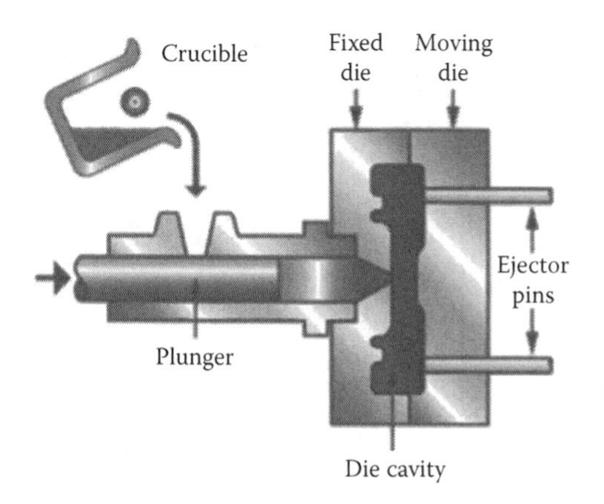

Figure 6.3 High-pressure die-casting.

tooling is a major concern since they are reused over and over, sometimes for hundreds of thousands of parts (Raman and Wadke 2006). Thermal management of tooling through spraying and cooling channels is also imperative since thermal fatigue is a major failure mode for this type of tooling (Sirinterlikci 2000).

Common materials that are cast include metals such as aluminum, magnesium, copper, and their alloys, low-melting-point alloys including zinc alloys, cast iron, and steel (Raman and Wadke 2006). The tooling used have simple ways of introducing liquid metal, feeding it into the cavity, and have mechanisms to exhaust air or gas entrapped within as well as to prevent defects such as shrinkage porosity and promote solid castings and easy removal of the part or parts. Cores and cooling channels are also included for making voids in the parts and controlled cooling of them to reduce cycle times for the process, respectively. The die or mold design, metal fluidity, and solidification patterns are all critical to obtain high-quality castings. Suitable provisions are made through allowances to compensate for shrinkage and finishing (Raman and Wadke 2006).

Bulk forming processes

Forming processes include bulk metal forming as well as sheet-metal operations. No matter what the type or nature of process is forming is mainly applicable to metals that are workable by plastic deformation. This constraint makes brittle materials not eligible for forming. Bulk forming is the combined application of temperature and pressure to modify shape of a solid object (Raman and Wadke 2006). While cold forming processes conducted near room temperature require higher pressures, hot working processes take advantage of the decrease in material strength. Consequent pressure and energy requirements are also much lower in hot working, especially when the material is heated above its recrystallization temperature, 60% of the melting point (Raman and Wadke 2006). Net shape and near net shape processes accomplish part dimensions that are exact or close to specification requiring little or no secondary finishing operations.

The following group of operations can be included in the classification of bulk forming processes (Kalpakjian and Schmid 2006):

1. *Rolling processes*: Flat rolling, shape rolling, ring rolling, and roll forging
2. *Forging processes*: Open-die forging, closed-die forging, heading, and piercing
3. *Extrusion and drawing processes*: Direct extrusion, cold extrusion, drawing, and tube drawing

In the flat rolling process, two rolls rotating in opposite directions are utilized in reducing the thickness of a plate or sheet metal. This thickness reduction is compensated for by an increase in the length, and when the thickness and width are close in dimension, an increase in both width and length occurs based on the preservation of the volume of the parts (Raman and Wadke 2006). A similar process, shape rolling (Figure 6.4) is used for obtaining different shapes or cross sections. Forging is used for shaping objects in a press and additional tooling, and may involve more than one pre-forming operation, including blocking, edging, and fullering (Raman and Wadke). Open-die forging is done on a flat anvil and closed-die forging process (Figure 6.5) uses a die with a distinct cavity for shaping. Open-die forging is less accurate but can be used in making extremely large parts due to its ease on pressure and consequent power requirements. Mechanical hammers deliver sudden loads whereas hydraulic presses apply gradually increasing loads. Swaging is a rotary

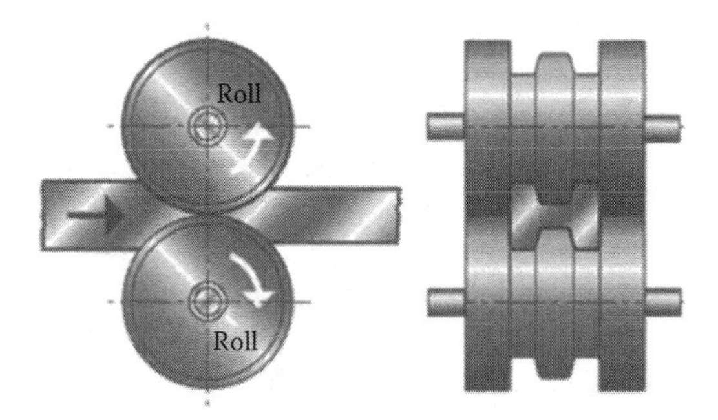

Figure 6.4 Shape rolling process.

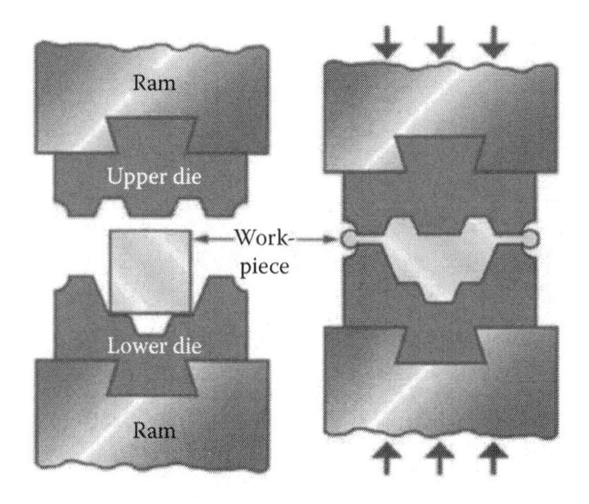

Figure 6.5 Open-die forging.

variation of the forging process, where a diameter of a wire is reduced by reciprocating movement of one or two opposing dies (Figure 6.6). Extrusion is the forcing of a billet out of a die opening similar to squeezing toothpaste out of its container (Figure 6.7), either directly or indirectly. This process enables fabrication of different cross sections on long pieces (Raman and Wadke 2006). In co-extrusion, two different materials are extruded at the same time and bond with each other. On the contrary, drawing process is based on pulling of a material through an orifice to reduce the diameter of the material.

Sheet forming processes

Stamping is a generic term used for sheet-metal processes. They include processes such as blanking, punching (piercing), bending, stretching, deep drawing, bending, and coining (Figure 6.8). Stamping processes are

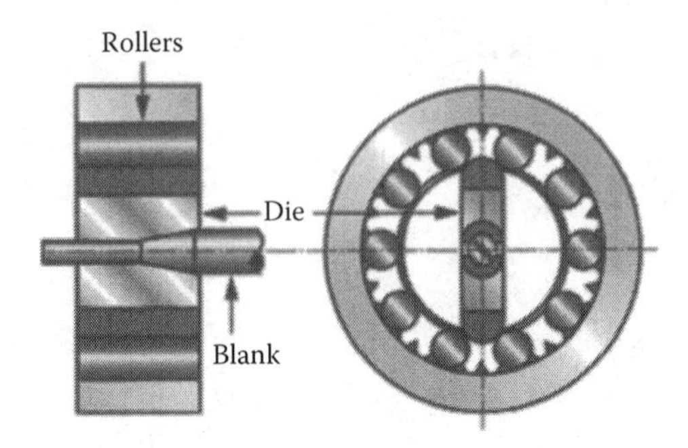

Figure 6.6 Swaging.

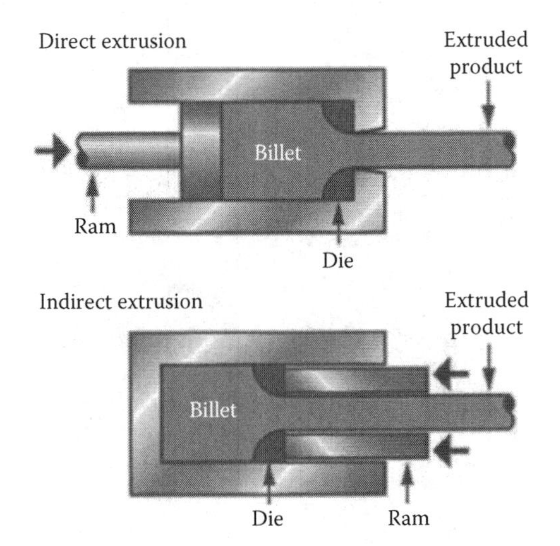

Figure 6.7 Extrusion.

executed singularly or consecutively to obtain complex sheet-metal parts with a uniform sheet-metal thickness. Progressive dies allow multiple operations to be performed at the same station. Since these tooling elements are dedicated, their costs are high and expected to perform without failure for the span of the production.

Sheet-metal pieces are profiled by a number of processes based on shear fracture of the sheet metal. In punching, a circular or a shaped hole is obtained by pushing the hardened die (punch) through the sheet metal (Figure 6.9). In a similar process called perforating, a group of punches are employed in making a hole pattern. In the blanking process, the aim is to keep the part that is punched out by the punch, not the rest of the sheet metal with the punched hole in the punching process (Figure 6.9). In the nibbling process, a sheet supported by an anvil, is cut to a shape by successive bites of a punch similar to the motion of the sewing machine head (Figure 6.9).

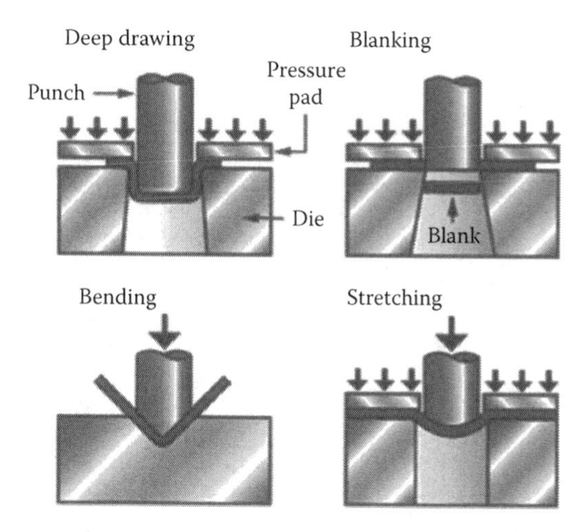

Figure 6.8 Stamping processes.

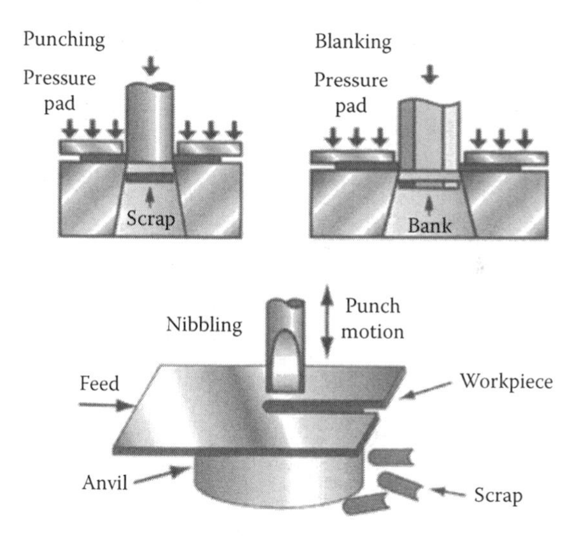

Figure 6.9 Punching, blanking, and nibbling processes.

After the perforation of a sheet-metal part, multiple different stamping operations may be applied to it. In simple bending, the punch bends the blank on a die (Figure 6.8). Stretching may be accomplished while strictly holding the sheet-metal piece by the pressure pads and forming it in a die, whereas the sheet-metal piece is allowed to be deeply drawn into the die while being held by the pressure pads in deep drawing (Figure 6.8). More sophisticated geometries can be obtained by roll forming (Figure 6.10) in a progressive setting.

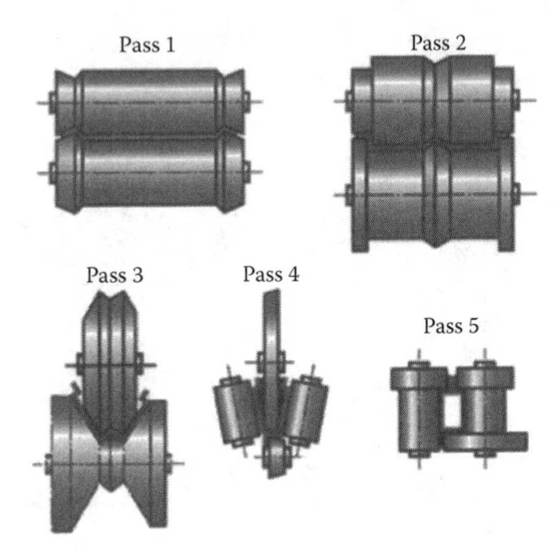

Figure 6.10 Roll forming of sheet metal.

Polymer processes

Polymer extrusion is a process to make semi-finished polymer products such as rods, tubes, sheets, and film in mass quantities. Raw materials such as pellets or beads are fed into the barrel to be heated and extruded at temperatures as high as 370°C (698°F) (Figure 6.11). The extrusion is then air or water cooled and may later be drawn into smaller cross sections. Variations of this process are film blowing, extrusion blow molding, and filament forming. This extrusion process is used in making of blended polymer pellets and becomes a post process for other processes such as injection molding. It is also utilized in coating metal wires in high speeds.

Injection molding (Figure 6.12) is a process similar to polymer extrusion with one main difference, the extrusion being forced into a metal mold for solidification under pressure and cooling. Feeding system into the mold includes the sprue area, runners, and gate. Thermoplastics,

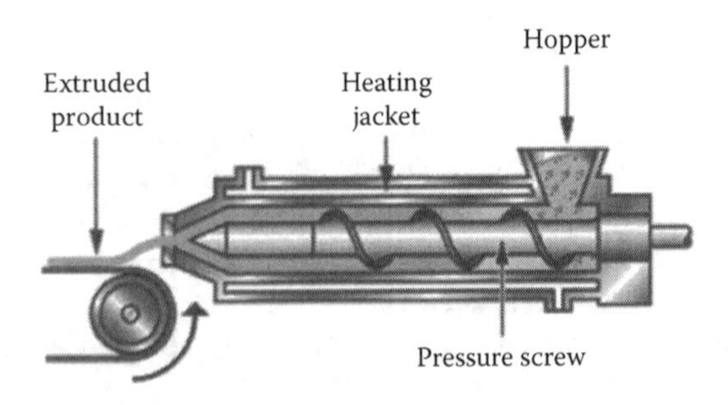

Figure 6.11 Polymer extrusion.

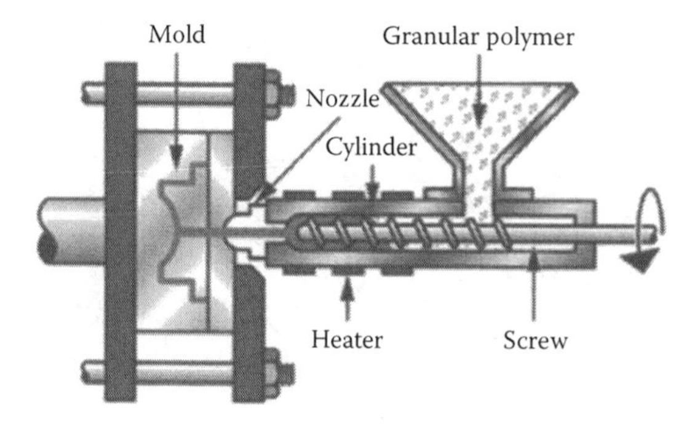

Figure 6.12 Injection molding.

thermosets, elastomers, and even metal materials are being injection molded. Co-injection process allows molding of parts with different material properties including colors and features. While injection foam molding with inert gas or chemical blowing agents results in making of large parts with solid skin and cellular internal structure, reaction injection molding (RIM) mixes low-viscosity chemicals under low pressures (0.30–0.70 MPa) (43.51–101.53 psi) to be polymerized via a chemical reaction inside a mold (Figure 6.13). The RIM process can produce complex geometries and works with thermosets such as polyurethane or other polymers, such as nylons, and epoxy resins. The RIM process is also adapted to fabricate fiber-reinforced composites.

Adapted from glass blowing technology, blow molding process utilizes hot air to push the polymer against the mold walls to be frozen (Figure 6.14). The process has multiple variations including extrusion and stretch blow molding. The generic blow molding process allows inclusion

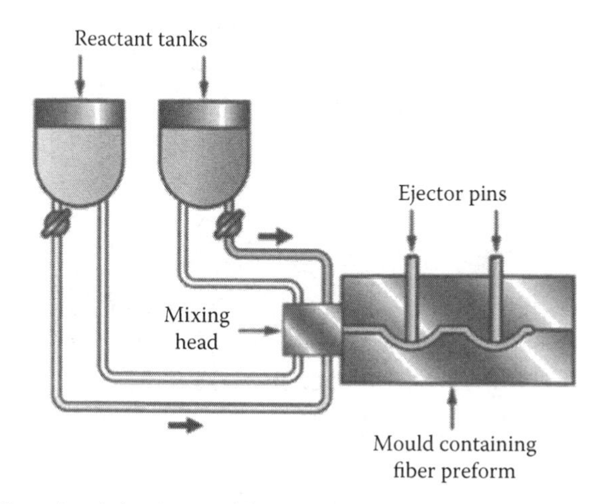

Figure 6.13 Reaction injection molding.

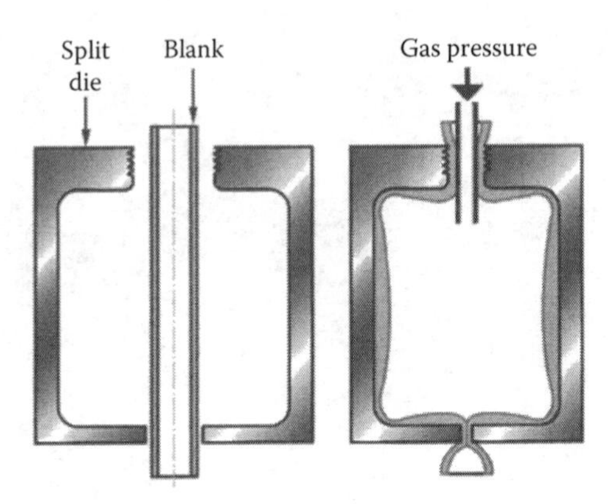

Figure 6.14 Blow molding.

of solid handles and has better control over the wall thickness compared with its extrusion variant.

Thermoforming processes are used in making large sheet-based moldings (Figure 6.15). Vacuum thermoforming applies vacuum to draw the heated and softened sheet into the mold surface to form the part. Drape thermoforming take advantages of the natural sagging of the heated sheet in addition to the vacuum whereas plug-assisted variant of thermoforming supplements the vacuum with a plug by pressing on the sheet. In addition, pressure thermoforming applies a few atmospheres (atm) to push the heated sheet into the mold. Various molding materials employed in the thermoforming processes included wood, metal, and polymer foam.

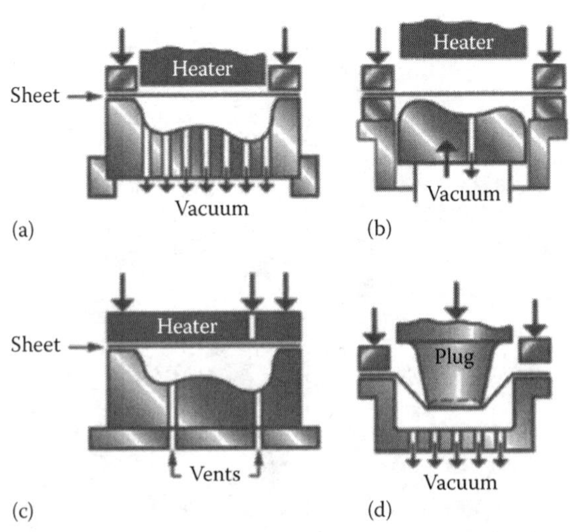

Figure 6.15 Thermoforming: (a) vacuum thermoforming, (b) drape thermoforming, (c) pressure thermoforming, and (d) plug-assisted thermoforming.

A wide variety of other polymer processing methods are available including but not limited to rotational molding, compression molding, and resin transfer molding (RTM).

Machining processes

Machining processes use a cutting tool to remove material from the workpiece in the form of chips (Raman and Wadke 2006). The cutting process requires plastic deformation and consequent fracture of the workpiece material. The type of chip impacts both the removal of the material and the quality of surface generated. The size and the type of the chip are a dependent of the type of machining operation and cutting parameters. The chip types are continuous, discontinuous, continuous with a built-up edge, and serrated (Raman and Wadke 2006). The critical cutting parameters include the cutting speed (in revolutions per minute—rpm's, surface feet per minute [sfpm] or millimeters per minute [mm/ min]), the feed rate (inches per minute [ipm] or millimeters per minute [mm/min]), and the depth of cut (inches [in] or millimeters [mm]). These parameters affect the workpiece, the tool, and the process itself (Raman and Wadke 2006). The conditions of the forces, stresses, and temperatures of the cutting tool are determined by these parameters. Typically, the workpiece or tool are rotated or translated such that there is relative motion between the two. A primary zone of deformation causes shear of material separating a chip from the workpiece. A secondary zone is also developed based on the friction between the chip and cutting tool (Raman and Wadke 2006). While rough machining is an initial process to obtain the desired geometry without accurate dimensions and surface finish, finish machining is a precision process capable of great dimensional accuracy and surface finish. Besides metals, stones, bricks, wood, and plastics can be machined.

There are usually three types of chip-removal operations: single-point, multipoint (fixed geometry), and multipoint (random geometry) (Raman and Wadke 2006). Random geometry multipoint operation is also referred to as abrasive machining process, and includes operations such as grinding (Figure 6.16), honing, and lapping. The cutting tool in a single-point operation resembles a wedge, with several angles and radii to aid cutting. The cutting-tool geometry is characterized by the rake angle, lead or main cutting-edge angle, nose radius, and edge radius. Common single-point operations include turning, boring, and facing (Raman and Wadke 2006). Turning is performed to make round parts (Figure 6.17), facing makes flat features, and boring fabricates nonstandard diameters, and internal cylindrical surfaces. Multipoint (fixed geometry) operations include milling and drilling. Milling operations (Figure 6.17) can be categorized into face milling, peripheral (slab) milling, and end milling.

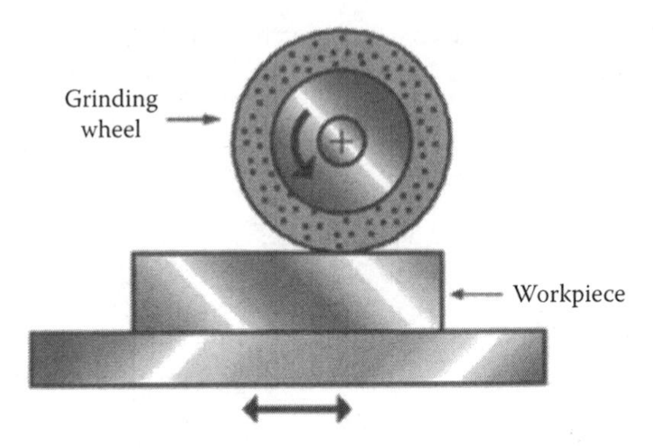

Figure 6.16 Grinding—used for finishing in general or machining hard materials.

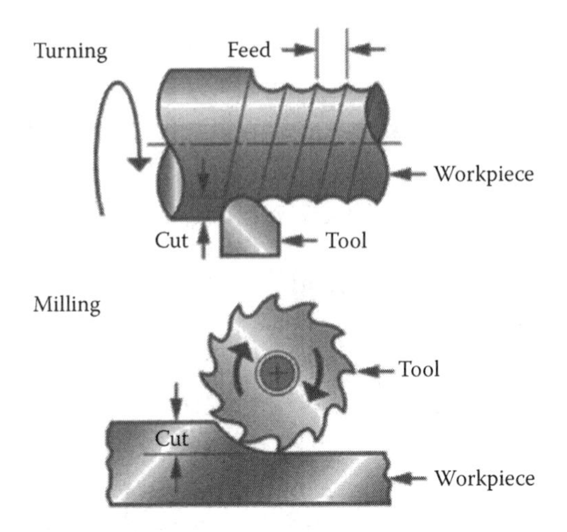

Figure 6.17 Machining processes.

The face milling uses the face of the tool, while slab milling uses the periphery of the cutter to generate the cutting action (Figure 6.18). These are typically applied to make flat features at a rate of material removal significantly higher than single-point operations like shaping and planning (Raman and Wadke 2006). End milling cuts along with both the face and periphery and is used for making slots and extensive contours (Figure 6.18). Drilling is used to make standard-sized holes with a cutter with multiple active cutting edges (flutes). Rotary end of the cutter is used in the material removal process. Drilling has been the fastest and most economical method of making holes into a solid object. A multitude of drilling and relevant operations are available including core drilling, step (peck) drilling, counterboring and countersinking as well as reaming and tapping (Figure 6.19).

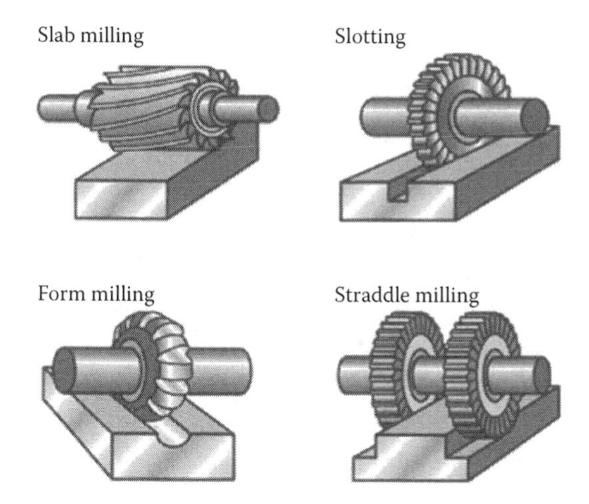

Figure 6.18 Milling operations.

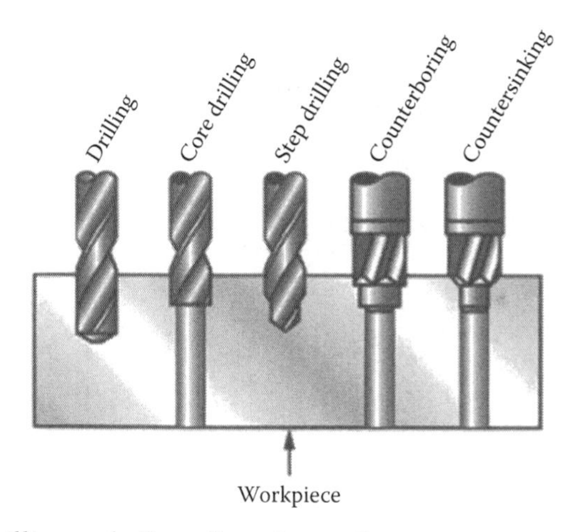

Figure 6.19 Drilling and other relevant operations.

Assembly and joining processes

Joining processes are employed in the manufacture of multipiece parts and assemblies (Raman and Wadke 2006). These processes encompass mechanical fastening through removable bolting as shown in Figure 6.20 and nonremoval riveting as shown in Figure 6.21, adhesive bonding, and welding processes.

Welding processes use different heat sources to cause localized melting of the metal parts to be joined or the melting of a filler to develop a joint between mainly two metals—also being heated. Welding of plastics has also been established. Cleaned surfaces are joined together through

Figure 6.20 Threaded fasteners.

Figure 6.21 Rivets and staples.

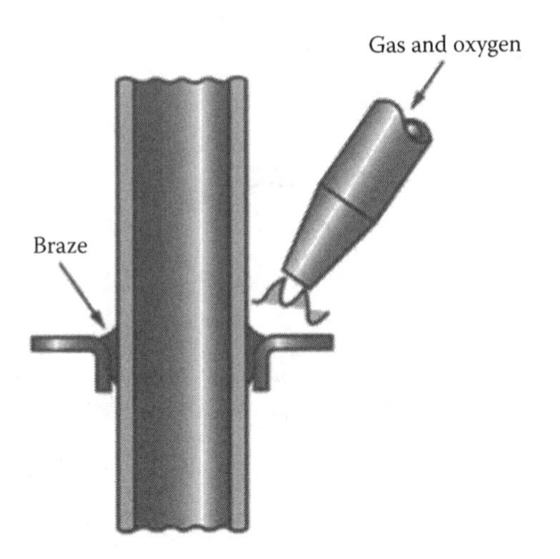

Figure 6.22 Brazing.

a butt weld or a lap weld, although other configurations are also feasible (Raman and Wadke 2006). Two other joining processes are brazing (Figure 6.22) and soldering, which differ from each other in the process temperatures and are not as strong as welding (Figure 6.22).

Arc welding utilizes an electric arc between two electrodes to generate the required heat for the process. One electrode is the plate to be joined while the other electrode is the consumable one. Stick welding is most common, and is also called shielded metal arc welding (SMAW) (Figure 6.23). Metal inert gas (MIG) or gas metal arc welding (Figure 6.24) uses a consumable electrode (Raman and Wadke 2006) as well. The electrode provides the filler and the inert gas provides an atmosphere such that contamination of the weld pool is prevented and consequent weld quality is obtained. A steady flow of electrode is accomplished through

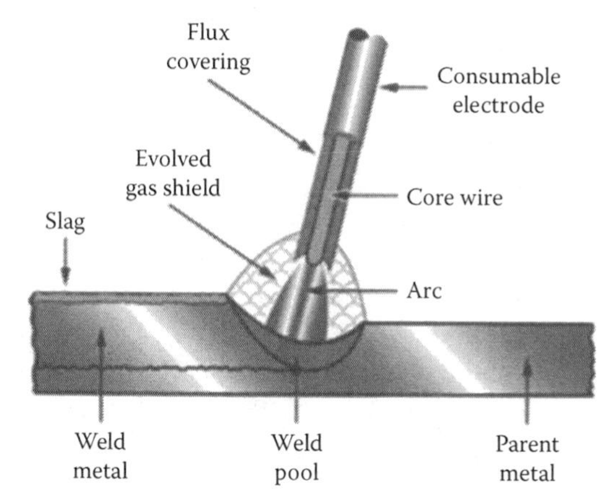

Figure 6.23 Stick (SMAW) welding.

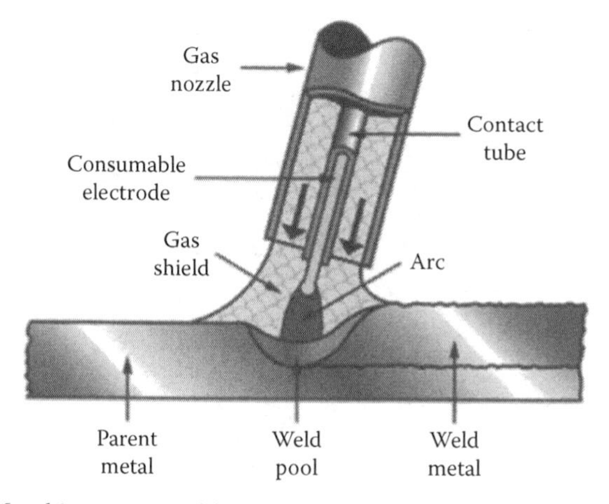

Figure 6.24 Metal inert gas welding.

automatically to maintain the arc gap, sequentially controlling the temperature of the arc (Raman and Wadke 2006). Gas tungsten arc welding or tungsten inert gas (TIG) welding uses a nonconsumable electrode, and filler is required for the welding (Figure 6.25). In resistance welding, the resistance is generated by the air gap between the surfaces to obtain and maintain the flow of electric current between two fixed electrodes. The electrical current is then used to generate the heat required for welding (Raman and Wadke 2006). This process results in spot or seam welds. Gas welding (Figure 6.26) typically employs acetylene (fuel) and oxygen (catalyzer) to develop different temperatures to heat workpieces or fillers for welding, brazing, and soldering. If the acetylene is in excess, a reducing (carbonizing) flame is obtained. The reducing flame is used in hard-facing

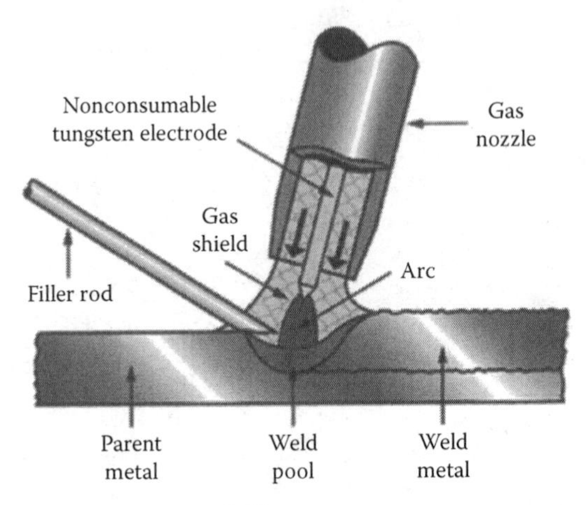

Figure 6.25 Tungsten inert gas welding.

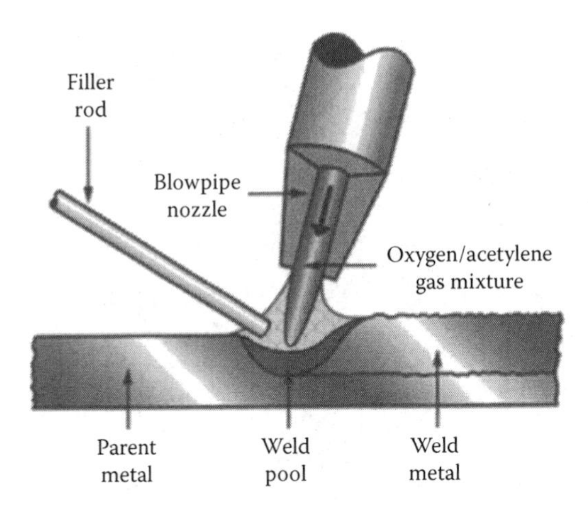

Figure 6.26 Oxy-fuel welding.

or backhand pipe welding operations. On the contrary, if oxygen is in excess, then an oxidizing flame is generated. The oxidizing flame is used in braze-welding and welding of brasses or bronzes. Finally, if equal proportions of the two are used, a neutral flame results. The neutral flame is used in welding or cutting. Other solid-state processes include thermit welding, ultrasonic welding, and friction welding.

Finishing processes

Finishing processes include surface treatment processes and material removal processes such as polishing, shot-peening, sand blasting, cladding and electroplating (Figure 6.27), and coating and painting (Raman and

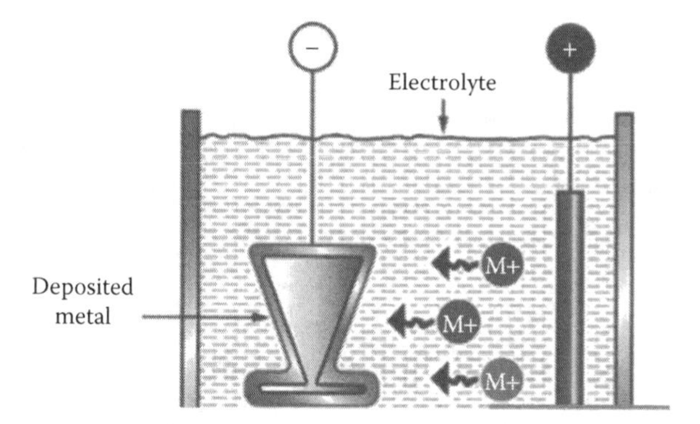

Figure 6.27 Electroplating.

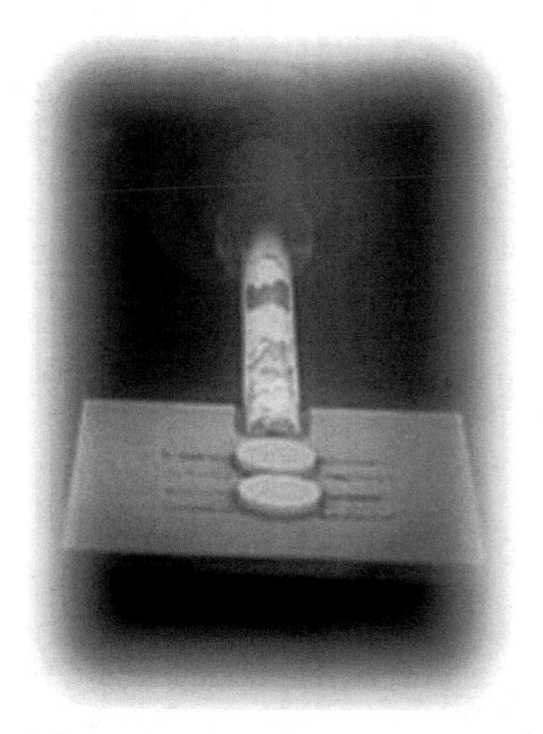

Figure 6.28 Physical vapor deposition.

Wadke 2006). Polishing involves very little material removal and is also classified under machining operations. Shot and sand blasting are used to improve surface properties and cleanliness of parts. Chemical vapor deposition (CVD) or physical vapor deposition (PVD) (Figure 6.28) methods are also applied to improve surface properties (Kalpakjian and Schmid 2006). Hard coatings such as CVD or PVD are applied to softer substrates to improve wear resistance while retaining fracture resistance (Raman and Wadke 2006) as in the case metal-coated polymer injection molding inserts (Noorani 2006). The coatings are less than 10 mm thick in many cases. On the other hand, cladding is done as in the case of aluminum cladding on stainless steel to improve its heat conductivity for thermally critical applications (Raman and Wadke 2006).

Nontraditional manufacturing processes

There are many nontraditional manufacturing processes. Nontraditional processes include the following:

1. *Electrically based machining*: These processes include the EDM (Figure 6.29), and electrochemical machining (ECM). In the Plunge EDM process, the workpiece is held in a work-holder submerged in a dielectric fluid. Rapid electric pulses are discharged between the graphite electrode and the workpiece, causing plasma to erode the workpiece. The dielectric fluid then carries the debris. The wire EDM uses mainly brass wire in place of the graphite electrode, but functions in a similar way (Figure 6.30). The ECM (also called reverse electroplating) process is similar to the EDM processes but it does not cause any tool wear, nor can any sparks be seen. Both processes can be used in machining very hard materials that are electrically conductive.
2. *Laser machining*: Lasers are used in a variety of applications ranging from cutting of complex 3D contours (i.e., today's coronary stents) to

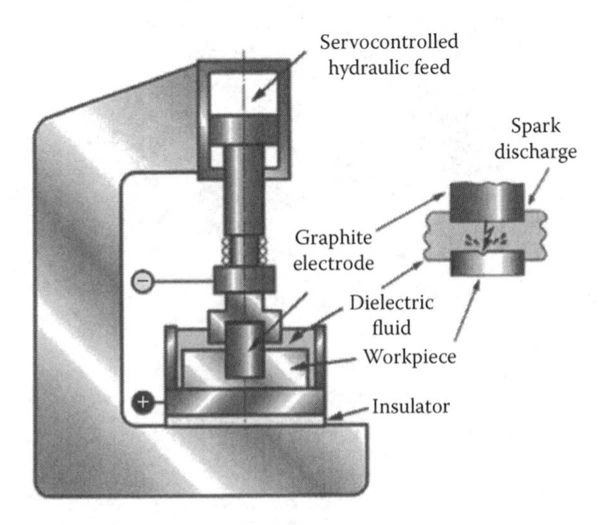

Figure 6.29 Electro-discharge machining.

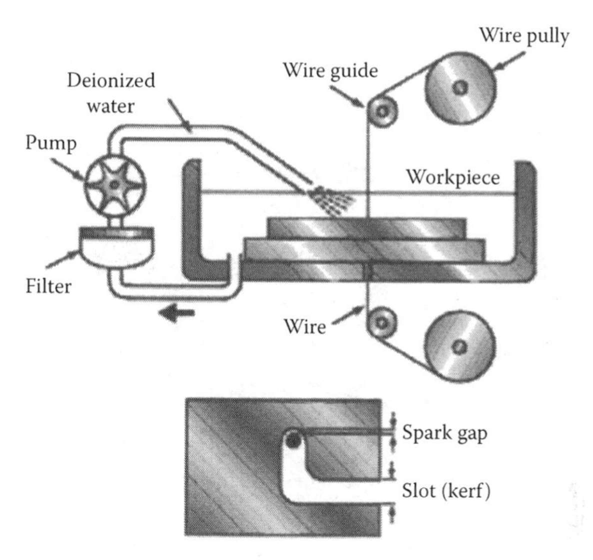

Figure 6.30 Wire EDM.

etching or engraving patterns on rolls for making texture on rolled parts. Lasers are also effectively used in hole-making, precision micromachining, removal of coating, and ablation. Laser transformation and shock hardening processes make workpiece surfaces very hard while the laser surface melting process produces refined and homogenized microstructures (Figure 6.31).

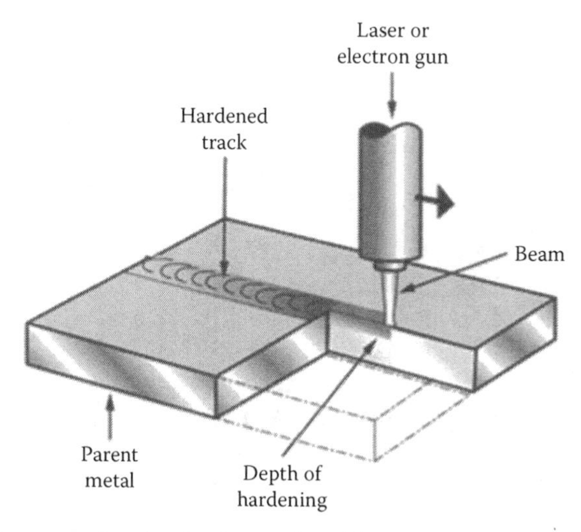

Figure 6.31 Laser surface hardening and melting.

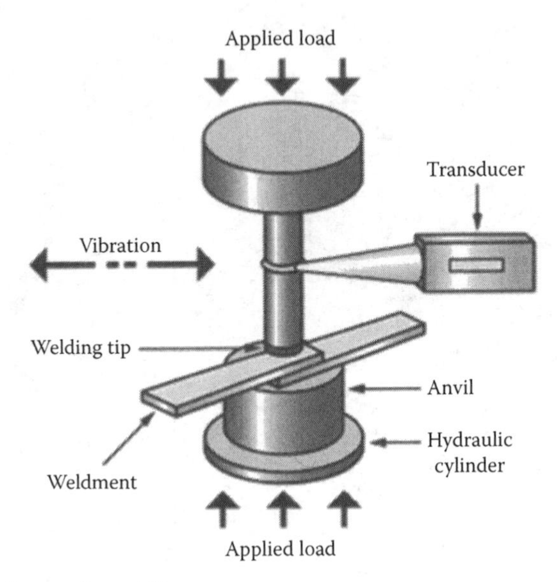

Figure 6.32 Ultrasonic welding.

3. *Ultrasonic welding*: This process requires an ultrasonic generator, a converter, a booster, and a welding tool (Figure 6.32). The generator converts 50 Hz into 20 KHz. These higher frequency signals are then transformed into mechanical oscillations through reverse piezoelectric effect. The booster and the welding tool transmit these oscillations into the welding area causing vibrations of 10–30 μm in amplitude. Meanwhile a static pressure of 2–15 MPa is applied to the workpieces as they slide, get heated, and bonded.
4. *Water-jet cutting*: This is an abrasive machining process employing abrasive slurry in a jet of water to machine hard to machine materials (Raman and Wadke 2006) (Figure 6.33). Water is pumped at high pressures like 400 MPa, and can reach to speeds of 850 m/s (3060 km/h or 1901.4 mph). Abrasive slurries are not needed when cutting softer materials.
5. *Powder metallurgy*: Powder-based fabrication methods are critical in employing materials with higher melting points due to hardship of casting them (Figure 6.34). Once compressed under pressures using different methods and temperatures compacted (green) powder parts are sintered (fused) usually at two-third of their melting points. Common powder metallurgy materials are ceramics and refractory metals, stainless steel, and aluminum.
6. *Small-scale manufacturing*: Last two decades have seen marriage of microscale electronics device and their manufacturing with mechanical systems—leading to the design and manufacturing of

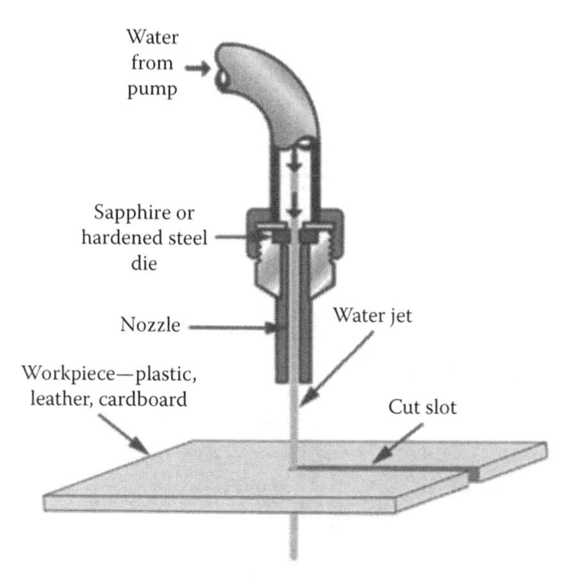

Figure 6.33 Water-jet cutting.

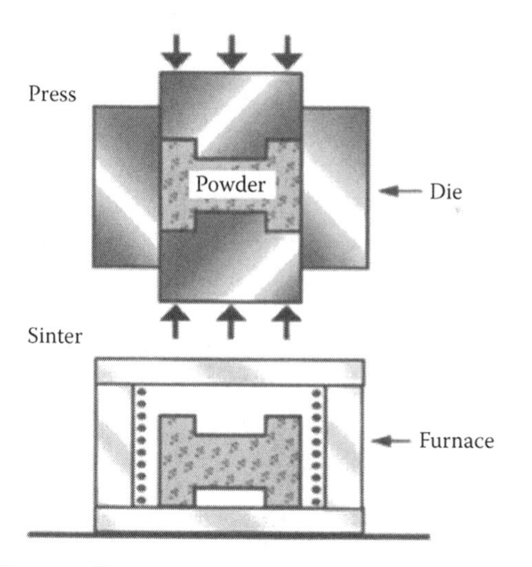

Figure 6.34 Powder metallurgy.

Micro-electromechanical devices (MEMS). Even newer cutting-edge technologies have emerged in a smaller scale such as nanotechnology or molecular manufacturing (Raman and Wadke 2006). The ability to modify and construct products at a molecular level makes nanotechnology very attractive and usable. Single-wall nanotubes are one of the biggest innovations for building future transistors and sensors. Variants of the nano area include nanomanufacturing, such as ultrahigh-precision machining or adding ionic antimicrobials to biomedical devices (Raman and Wadke 2006; Sirinterlikci et al. 2012).

7. *Additive manufacturing*: This process has been driven from additive rapid prototyping technology. Since late 1980s rapid prototyping technologies have intrigued scientists and engineers. In the early years, there were attempts of obtaining 3D geometries using various layered approaches as well as direct 3D geometry generation by robotic plasma spraying. Today, a few processes such as fused deposition modeling (FDM) (Figure 6.35), stereolithography (SLA), laser sintering processes (selective or direct metal) (Figure 6.36), and 3D printing have become household names. There are also other very promising processes such as objet's polyjet (inkjet) 3D printing technology. In the last two decades, the rapid prototyping technology has seen increase in number of materials available for processing, layer thicknesses has become less while control systems have improved to better the accuracy of the parts. The end result is shortened cycle times, better quality functional parts. In addition, there have been many successful applications of rapid tooling and manufacturing.
8. *Biomanufacturing*: This process may encompass biological and biomedical applications. Thus, manufacturing of human vaccinations may use biomass from plants while biofuels are extracted from corn or other crops. Hydraulic oils, printer ink-technology, paints, cleaners, and many other products are taking advantage of the developments in biomanufacturing (Sirinterlikci et al. 2010). On the other hand, biomanufacturing is working with nanotechnology, additive manufacturing, and other emerging technologies to improve the biomedical engineering field (Sirinterlikci et al. 2012).

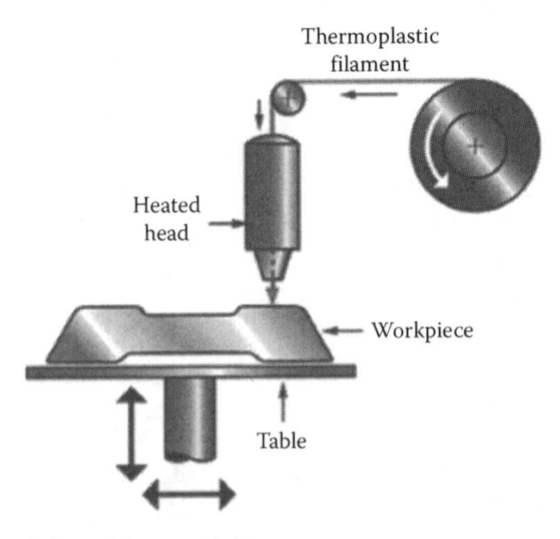

Figure 6.35 Fused deposition modeling.

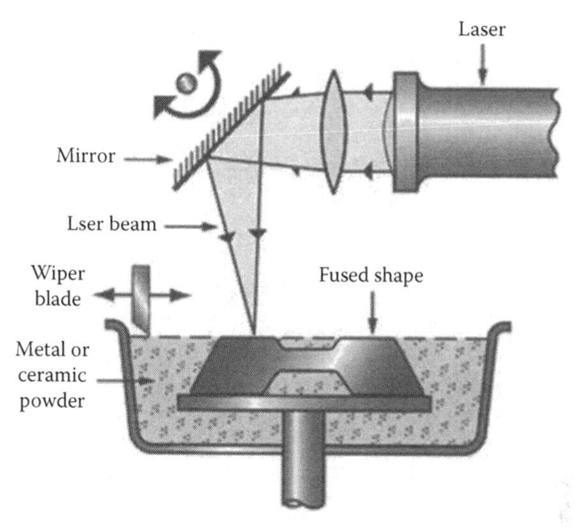

Figure 6.36 Selective laser sintering.

Manufacturing systems

The manufacturing systems are systems that are employed to manufacture products and the parts assembled into those parts (Groover 2001). A manufacturing system is the collection of equipment, people, information, processes, and procedures to realize the manufacturing targets of an enterprise. Groover (2001) defines manufacturing systems in the following two parts:

1. Physical facilities include the factory, the equipment within the factory, and the way the equipment is arranged (the layout).
2. Manufacturing support systems is a set of procedures of the company to manage its manufacturing activities and to solve technical and logistics problems including product design and some business functions.

Production quantity is a critical classifier of a manufacturing system while the way the system operates is another one including *production to order* (i.e., just-in-time—JIT manufacturing) and *production for stock* (Hitomi 1996). According to Groover (2001) and Hitomi (1996), a manufacturing system can be categorized into the following three groups based on its production volume:

1. *Low production volume—jobbing systems*: In the range of 1–100 units per year, associated with a job shop, fixed position layout, or process layout

2. *Medium production volume—intermittent or batch systems*: With the production range of 100–10,000 units annually, associated with a process layout or cell
3. *High production volume—mass manufacturing systems*: 10,000 units to millions per year, associated with a process or product layout (flow line)

Each classification is associated inversely with a product variety level. Thus, when product variety is high, production quantity is low, and when product variety is low, production quantity is high (Groover 2001). Hitomi (1996) states that only 15% of the manufacturing activities in late 1990s were coming from mass manufacturing systems while small-batch multi-product systems had a share more than 80%, perhaps due to diversification of human demands.

While each manufacturing process is important for fabrication of a single part, assembling these piece parts into subassemblies and assemblies also present a major challenge. The concept that makes assembly a reality is interchangeability. Interchangeability relies on standardization of products and processes (Raman and Wadke 2006). Besides facilitating easy assembly, interchangeability also enables easy and affordable replacement of parts within subassemblies and assemblies (Raman and Wadke 2006).

The key factors for interchangeability are *tolerances* and *allowances* (Raman and Wadke 2006). Tolerance is the permissible variation of geometric features and part dimensions on manufactured parts with the understanding that perfect parts are hard to be made, especially repeatedly. Even if the parts could be manufactured perfectly, current measurement tools and systems may not able to verify their dimensions and features accurately (Raman and Wadke 2006). Allowances determine the degree of looseness or tightness of a fit in an assembly of two parts (i.e., a shaft and its bearing) (Raman and Wadke 2006). Depending on the allowances, the fits are classified into *clearance, interference,* and *transition* fits. Since most commercial products and systems are based on assemblies, tolerances (dimensional or geometric) and allowances must be suitably specified to promote interchangeable manufacture (Raman and Wadke 2006).

Systems metrics and manufacturing competitiveness

According to Hitomi (1996), efficient and economical execution of manufacturing activities can be achieved by completely integrating the material flow (manufacturing processes and assembly), information flow (manufacturing management system), and cost flow (manufacturing economics). A manufacturing enterprise need to serve for the welfare of the society by not harming the people and the environment (green manufacturing, environmentally conscious manufacturing) along with targeting its profit objectives (Hitomi 1996). A manufacturing enterprise need to remain competitive and thus has to evaluate its products' values and/or the effectiveness of their manufacturing system by using the following three metrics (Hitomi 1996):

1. Function and quality of products
2. Production costs and product prices
3. Production quantities (productivity) and on-time deliveries
4. Following any industry regulations

A variety of means are needed to support these metrics including process planning and control; quality control; costing; safety, health, and environment (SHE); production planning; scheduling; and control including assurance of desired cycle/task and throughput times. Many other metrics are also used in the detail design and execution of systems including machine utilization, floor space utilization, and inventory turnover rates.

Automation of systems

Some parts of a manufacturing system needs to be automated, while other parts remain under manual or clerical control (Groover 2001). Either the actual physical manufacturing system or its support system can be automated as long as the cost of automation is justified. If both systems are automated a high level of integration can also be reached as in the case of computer-integrated manufacturing (CIM) systems. Groover (2001) lists examples of automation in a manufacturing system:

- Automated machines
- Transfer lines that perform multiple operations
- Automated assembly operations
- Robotic manufacturing and assembly operations
- Automated material handling and storage systems
- Automated inspection stations

Automated manufacturing systems are classified into the following three groups (Groover 2001):

1. *Fixed automation systems*: Used for high production rates or volumes and very little product variety as in the case of a welding fixture used in making circular welds around pressure vessels.
2. *Programmable automation systems*: Used for batch production with small volumes and high variety as in the case of a robotic welding cell.

3. *Flexible automation systems*: Medium production rates and varieties can be covered as in the case of flexible manufacturing cells with almost no lost time for changeover from one part family to another one.

Groover lists the following reasons for automating a manufacturing entity:

1. To increase labor productivity and costs by substituting human workers with automated means
2. To mitigate skilled labor shortages as in welding and machining
3. To reduce routine and boring manual and clerical tasks
4. To improve worker safety by removing them from the point of operation in dangerous tasks such nuclear, chemical, or high energy
5. To improve product quality and repeatability
6. To reduce manufacturing lead times
7. To accomplish processes that cannot be done manually
8. To avoid the high cost of not automating

Conclusions

Manufacturing is the livelihood and future of every nation and is mainly misunderstood. It also has a great role in driving engineering research and development; in addition to wealth generation. Utilization of efficient and effective methods is crucial for any manufacturing enterprise to remain competitive in this very global market, with intense international collaboration and rivalry. Especially the importance of industrial engineering tools in optimizing processes and integrating those processes into systems need to be grasped to better the manufacturing processes and their systems. Automation is still a valid medium for improving the manufacturing enterprise as long as the costs of doing it are justified. Tasks too difficult to automate, life cycles that are too short, products that are very customized, and cases where demands are very unpredictably varying, cannot justify application of automation (Hitomi 1996).

References

Dano, S., *Industrial Production Models: A Theoretical Study*, Springer, Vienna, Austria, 1966.

Gallager, P., *Presentation at the National Network of Manufacturing Innovation Meeting II*, Cuyahoga Community College, Cleveland, OH, July 9, 2012.

Groover, M., *Automation, Production Systems, and Computer-Integrated Manufacturing*, Prentice Hall, Upper Saddle River, NJ, 2001.

Harrington, J. Jr., *Computer Integrated Manufacturing*, Industrial Press, New York, 1973.

Hitomi, K., Moving toward manufacturing excellence for future production perspectives, *Industrial Engineering*, Vol. 26, No. 6, p. 32, 1994.

Hitomi, K., *Manufacturing Systems Engineering: A Unified Approach to Manufacturing Technology, Production Management, and Industrial Economics*, 2nd Edition, Taylor & Francis, Bristol, PA, 1996.

Kalpakjian, S., S. R. Schmid, *Manufacturing Engineering and Technology*, 5th Edition, Pearson Prentice Hall, Upper Saddle River, NJ, 2006.

Noorani, R., *Rapid Prototyping and Applications*, John Wiley & Sons, Hoboken, NJ, 2006.

Raman, S., A. Wadke, *Handbook of Industrial and Systems Engineering*, CRC Press, Boca Raton, FL, 2006.

Sirinterlikci, A., *Thermal Management and Prediction of Heat Checking in H-13 Die-Casting Dies*, PhD dissertation, The Ohio State University, Columbus, OH, 2000.

Sirinterlikci, A., C. Acors, A. Pogel, J. Wissinger, M. Jimenez, Antimicrobial technologies in design and manufacturing of medical devices, *SME Nanomanufacturing Conference*, Boston, MA, May 22–24, 2012.

Sirinterlikci, A., A. Karaman, O. Imamoglu, G. Buxton, P. Badger, B. Dress, Role of biomaterials in sustainability, *Proceedings of the 2nd Annual Conference of the Sustainable Enterprises of the Future*, Pittsburgh, PA, June 7–10, 2010.

Timms, H. L., M. F. Pohlen, *The Production Function in Business—Decision Systems for Production and Operations Management*, 3rd Edition, Irwin, Homewood, IL, 1970.

chapter seven

Manufacturing technology processes[*]

Manufacturing

The term *manufacturing* originates from the Latin word *manufactus*, which means *made by hand*. Manufacturing has seen several advances during the past three centuries: mechanization, automation and, most recently, computerization. Processes that used to be predominantly done by hand and hand tools have evolved into sophisticated processes making use of cutting-edge technology and machinery. A steady improvement in quality has resulted, with today's specifications even on simple toys exceeding those achievable just a few years ago. Mass production, a concept developed by Henry Ford, has advanced so much that it is now a complex, highly agile, and highly automated manufacturing enterprise.

From a technological standpoint, manufacturing involves the making of products from raw materials through the use of human labor and resources that include machines, tools, and facilities. It could be more generally regarded as the conversion of an unusable state into an usable state by adding value along the way. For instance, a log of wood serves as the raw material for making wood planks, which in turn are the raw material to produce chairs. The value added is usually represented in terms of cost and/or time.

Looking at manufacturing from a broad or systemic point of view, it envelops design, processes, controls, quality assurance, and all aspects related to the life cycle of the product from concept to product to recycling. Such a systemic view has been realized and improved over 75 years and yet the context and concept of integrating the various functions have become even more paramount today. Thus designs for manufacturability, assembly, quality, and environmentability are gathered under the general umbrella of *Design for X*. This is important considering that the preponderance of expenditures toward the development of a product occur in its initial stages and product design. Concurrent engineering

[*] Reprinted with permission from Adedeji B. Badiru (2014), *Handbook of Industrial and Systems Engineering*, Second Edition, Chapter 19 (S. Raman and A. Wadke), Pages 337–349, CRC Press/Taylor & Francis, Boca Raton, FL.

and simultaneous engineering are two other life cycle-based concepts that also promote integration rather than fragmentation. One reason for the systemic point-of-view stems from the demands placed on products today. It is not uncommon to find stringent tolerance and finish requirements on a small piece of lever that goes into a toy assembly that costs only a few dollars. Aesthetics, ergonomics, safety, and creativity each have experienced significant growth in the development of products.

Interchangeability

Interchangeable manufacture relies largely on standardization of products and processes. In the systemic sense, interchangeability facilitates *pluggability* or modularity and de-emphasizes design by product. Interchangeability must facilitate easy replacement of parts within assemblies at reasonable costs. The key words for interchangeability are *tolerances* and *allowances.* Tolerance is the permissible variation on manufactured parts with the understanding that perfect parts are impossible to make with currently available machines. Even if they could be manufactured perfectly, current measurement tools cannot suitably verify that. Instruments are limited by a least count or resolution, the best achievable with modern-day instruments being a nanometer. Allowances quantify the degree of looseness or tightness of a fit or assembly. Depending on the allowances, the fits are classified into *clearance, interference,* and *transition* fits. Since most commercial products including automobiles and airplanes are assemblies, tolerances and allowances must be suitably specified to promote interchangeable manufacture. Stated otherwise, if an assembly is completely composed of standard parts, that could each be procured *off-the-shelf,* the labor involved in designing dimensions on the piece parts can be minimized quite significantly. The final product will also have good longevity as part replacement will become simpler.

Manufacturing processes and process planning

Currently, several alternatives are available for processing and thus there is usually more than a certain way to manufacture any given part. The processes can be classified as follows:

1. Casting and foundry processes
2. Forming and shaping processes
3. Machining processes
4. Joining processes
5. Finishing processes
6. Nontraditional manufacturing processes

Every candidate process is suited to promote a particular functionality characteristic and is capable of generating a certain level of quality. Quality is a subjective metric, and often in the context of manufacturing implies surface finish and/or tolerances. The economics of any process is also very important and can be conveniently decomposed with the analysis of manufacturing operations. (Operations are subsets of processes and tasks are subsets of operations.) A manufactured part can be decomposed into several features, the features into operations, and the operations into tasks. Since several candidate operations may be selected and many sequences of operations exist, consequently, several viable process plans can be made.

Process planning is the coordinated selection and sequencing of manufacturing operations to manufacture a part. It is expected to provide detailed documentation of the human power and resources while rendering a product. Since the number of alternatives to manufacture a product may be very large, an exhaustive enumeration of all possible or feasible plans could be prohibitive; hence, a subset of the total number of viable sequences is derived from the grand list utilizing manufacturing precedence information. Objectives and constraints are identified and formulated suitably and weights drawn to evaluate and compare alternate plans. Considering the enormous effort involved, computer-aided process planning (CAPP) systems have become very attractive in order to generate feasible sequences and thereby to minimize the lead time and nonvalue-added costs.

Casting and foundry processes

Casting processes (Figures 7.1 and 7.2) can be further classified into permanent mold-type processes and expendable mold-type processes. The basic idea is to superheat a metal or alloy well beyond its melting point (often 3–5 times the melting point to increase the fluidity of the metal), pour it into a mold, and allow it to solidify within the mold. Upon solidification, the part is retrieved from the mold and finished suitably. The expendable mold processes destroy the mold after solidification and includes as examples sand casting, plaster molding, and investment casting. The last mentioned could result in better finishes and tighter tolerances than the former two. High-pressure and low-pressure die casting and centrifugal casting processes also result in good finishes, but are permanent mold processes. In these processes, the mold preservation is of principal concern since molds are reused for each part cast within them.

Common materials that are cast include metals such as aluminum, copper and their alloys, low melting point alloys, cast iron, and steels. The molds have a complex gating system with gates and runners that allow for

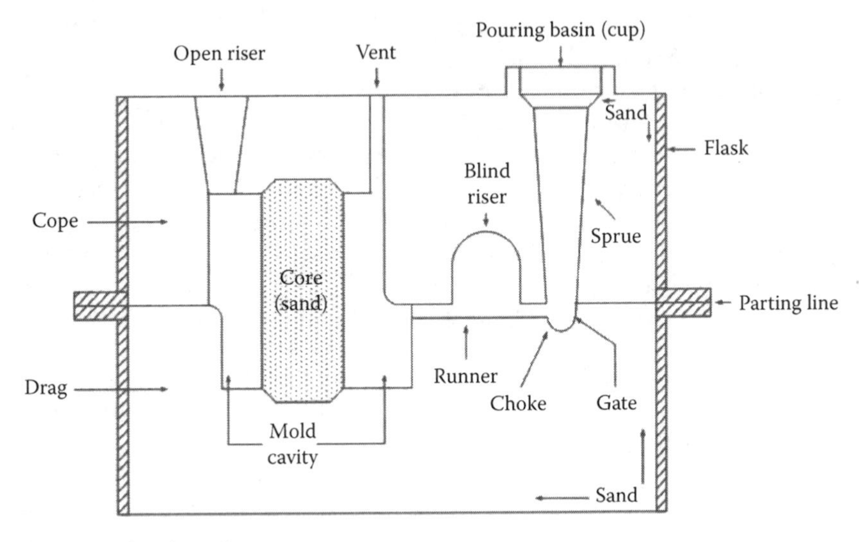

Figure 7.1 Sand casting.

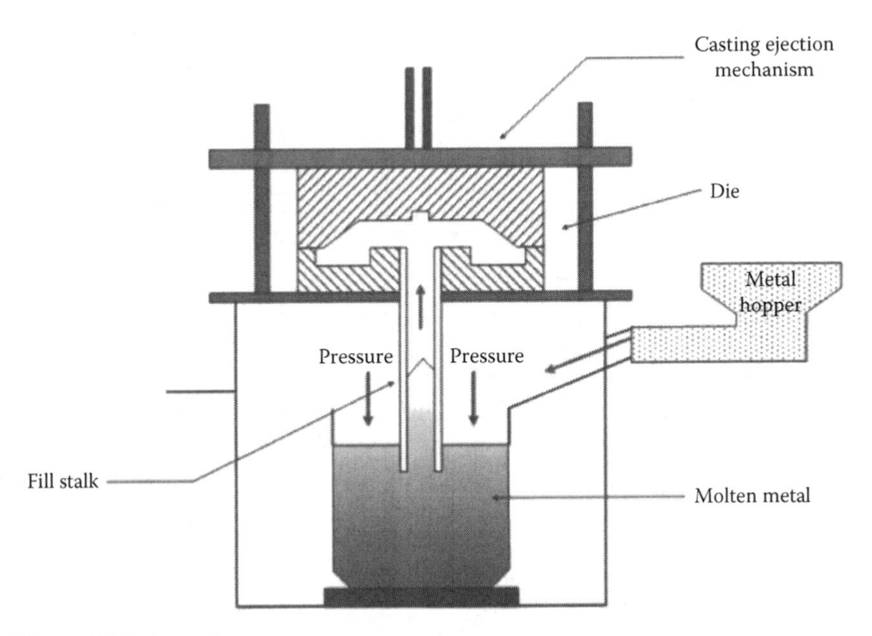

Figure 7.2 Die casting.

the delivery of liquid metal to the cavity, risers to compensate for shrinkage porosity, and a pouring cup. The mold design, metal fluidity, and solidification patterns are all very important to obtain high-quality castings devoid of defects. Suitable provisions are made through allowances to compensate for shrinkage and finishing.

Forming and working processes

Forming processes (Figures 7.3 through 7.6) include bulk metal forming as well as sheet-metal operations. Forming is largely applicable to metals that can be workable or malleable. Thus brittle materials are not suitable for forming. Forming is the combined application of temperature and pressure

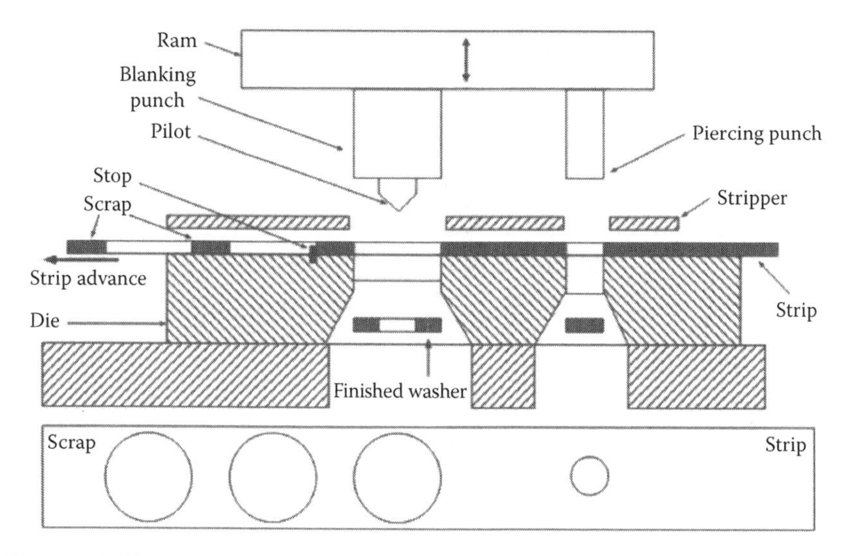

Figure 7.3 Shearing die.

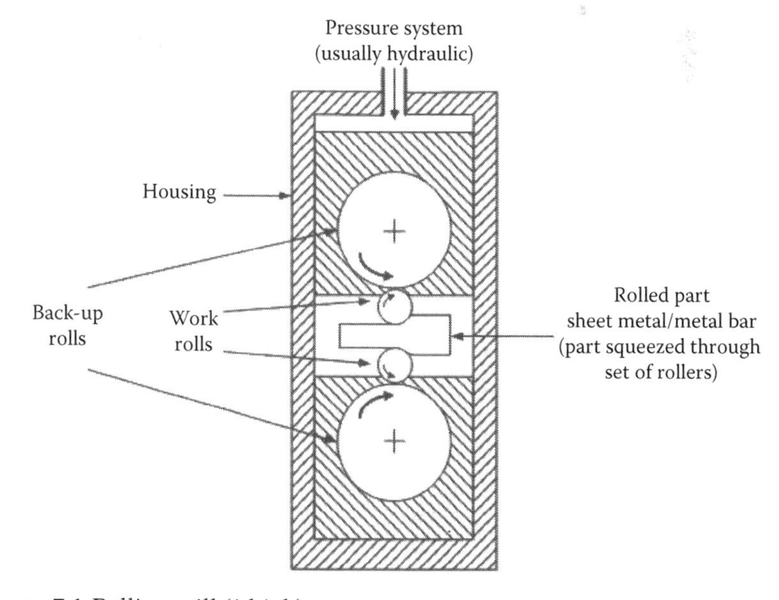

Figure 7.4 Rolling mill (4-high).

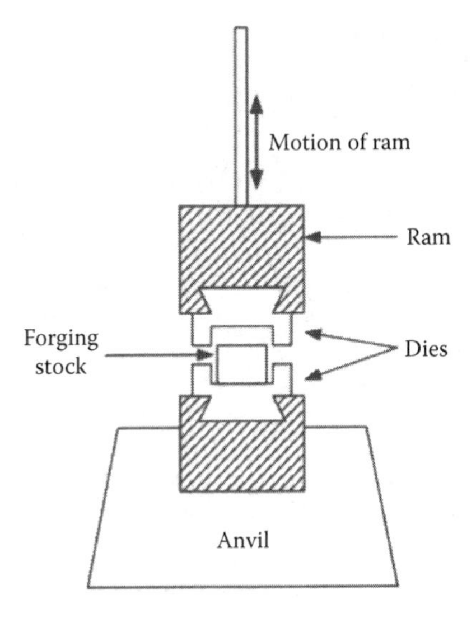

Figure 7.5 Forging.

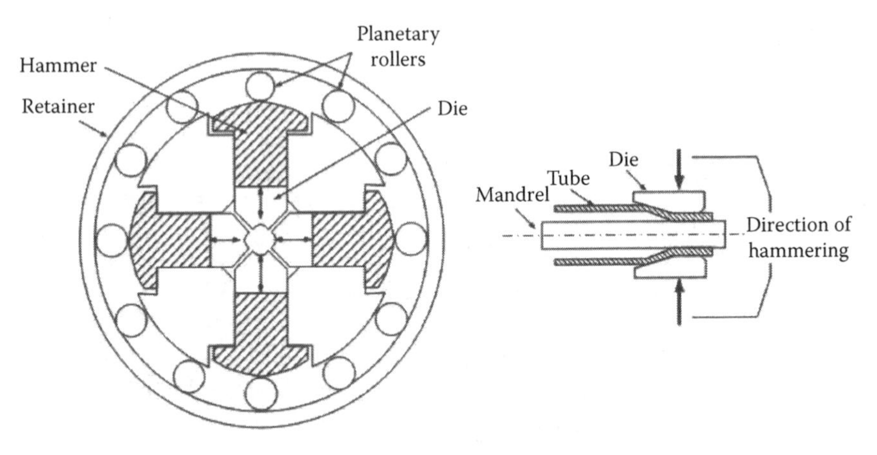

Figure 7.6 Swaging.

to shape an object to specifications in solid state. Cold forming includes operations performed close to the room temperature and consequently employs higher pressures to form. Hot working processes heat the workpart above its recrystallization temperature (60% of the melting point) and hence apply lower pressures to shape the object than used in cold working. A multitude of operations are classified within bulk forming processes of which rolling, forging, and extrusion garner the most attention.

Flat rolling is the process of reducing the thickness of a plate or sheet. Two rolls rotating in opposite directions scoop the workpart and reduce its thickness. This thickness reduction is typically compensated for by an increase in the length, and on occasion when the thickness and width are nearly the same an increase in both width and length occur. Shape rolling processes are used for obtaining different cross sections. Forging is used for shaping objects in general and may involve more than one pre-forming operation such as blocking, edging, and fullering. Open-die forging is done on a flat anvil and impression-die or closed-die forging processes use a die for shaping. Hammers which deliver instantaneous loads and presses that apply gradual loads are both used to supply the required pressures to shape the material. Swaging is a variation of the forging process. Coining is a net-shape forging operation that requires minimal post-processing. Extrusion is the pushing of a material through a die such as is common in squeezing toothpaste out of its dispenser. This allows for the fabrication of different cross sections on long rods. Drawing is similar with the material being pulled through a die rather than being pushed into it.

Sheet-metal operations include shearing, bending, stretch forming, spinning, and explosive forming. Shearing is usually the first operation performed in sheet-metal fabrication and can be categorized into blanking and piercing (punching). In the former operation, the slug cutout is important, whereas in the latter operation, the sheet from which the slug is cut is important. Sheet-metal forming utilizes progressive dies or compound dies such that multiple operations can be combined in producing a single component.

Machining processes

Machining processes (Figures 7.7 through 7.12), also called material removal processes, use a sharp tool to remove material from the workpiece in the form of chips. The process of cutting involves both plastic deformation and fracture. The type of chip that is generated during cutting is of high significance to the material removal as well as the quality of surface generated. The size and the type of the chip vary depending on the type of operation performed and the cutting parameters. Four types of chips are common in machining: continuous, discontinuous, continuous with a built-up edge, and serrated. The cutting parameters of importance include the cutting speed, the feed rate, and the depth of cut. These parameters affect the workpiece, the tool, and the process itself. The values of forces and stresses and temperatures on the cutting tool depend on these parameters. Typically, the workpiece and/or tool is rotated and translated such that there is relative motion between them. The cutting speed is the circumferential speed or the speed at which a new surface is being generated. A primary zone of deformation causes shear of material separating a chip from the workpiece. A secondary zone is developed due to the friction between the newly formed chip and the cutting tool.

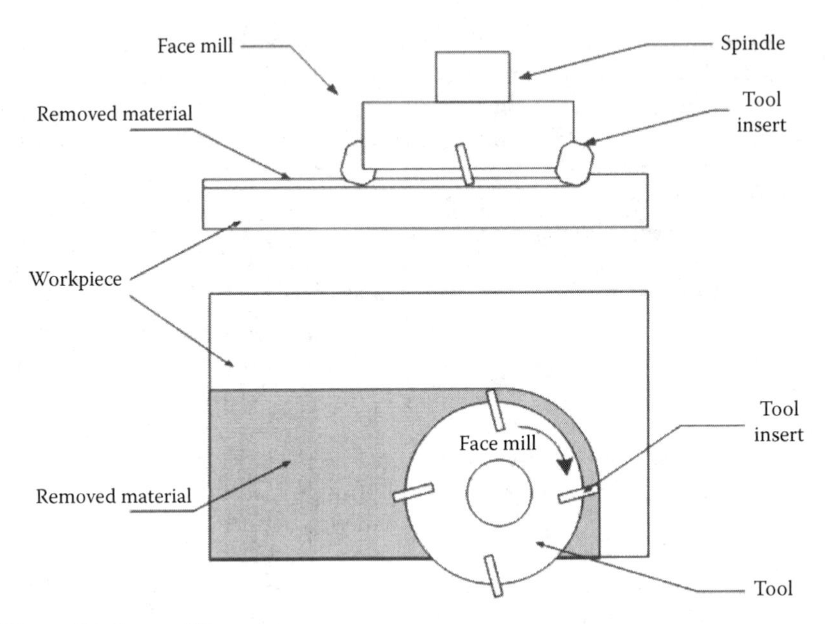

Figure 7.7 Face milling.

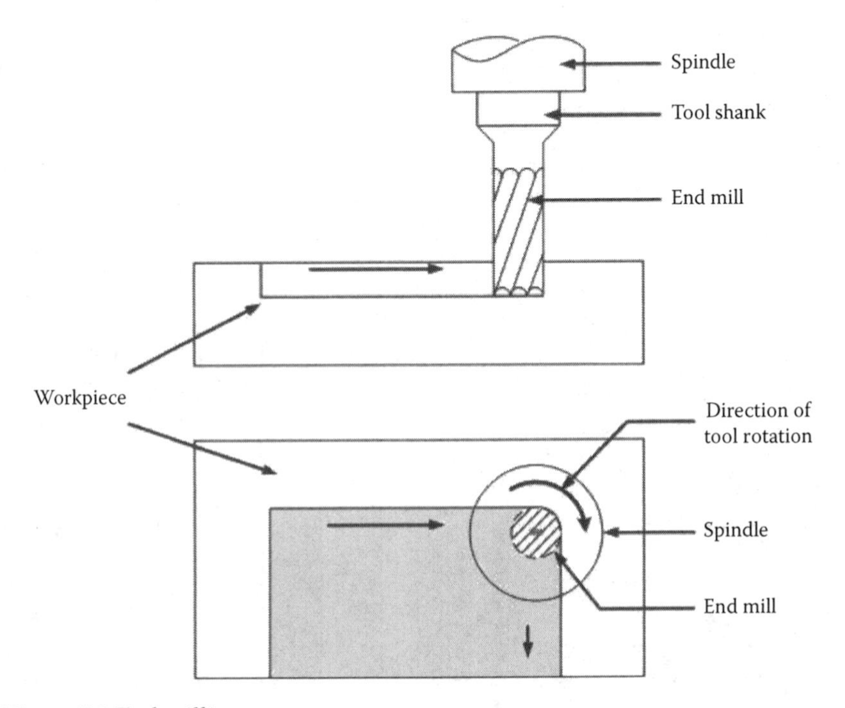

Figure 7.8 End milling.

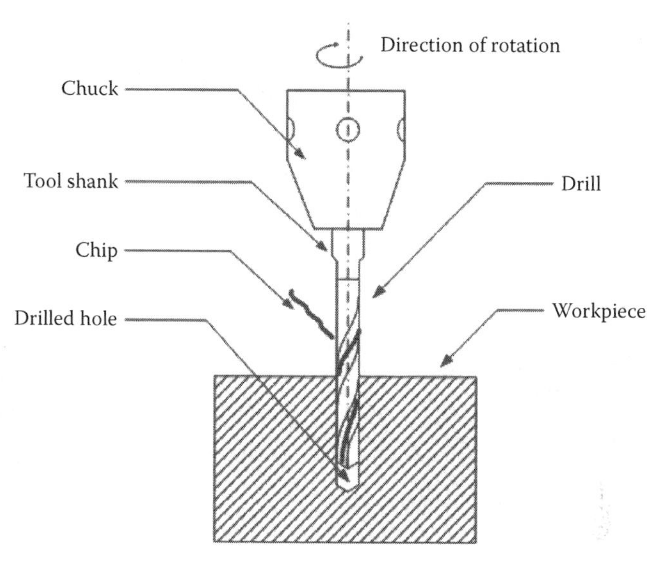

Figure 7.9 Drilling.

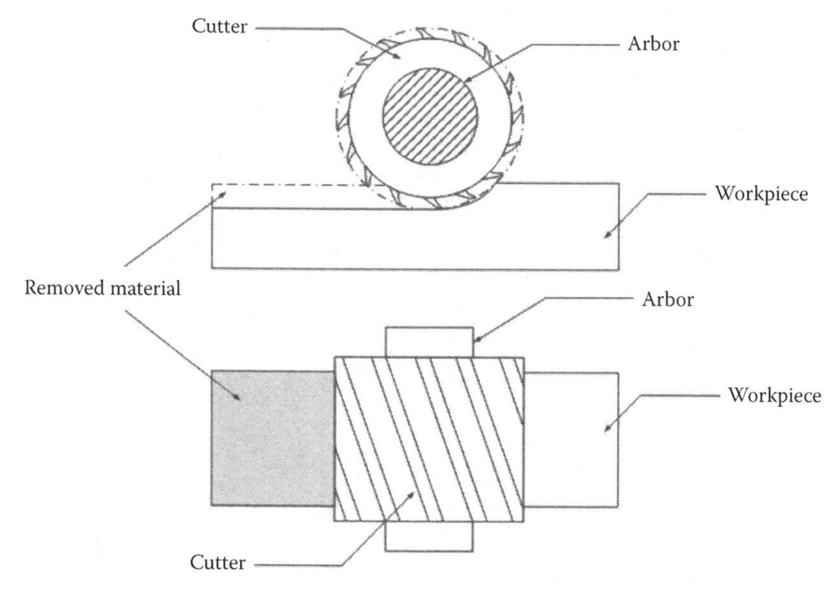

Figure 7.10 Slab milling.

There are usually three types of chip removal operations: single-point, multipoint (fixed geometry), and multipoint (random geometry). The latter is also termed *abrasive machining processes* and includes operations such as grinding, honing, and lapping. The cutting tool in a single-point operation resembles a wedge and is given several angles and radii to

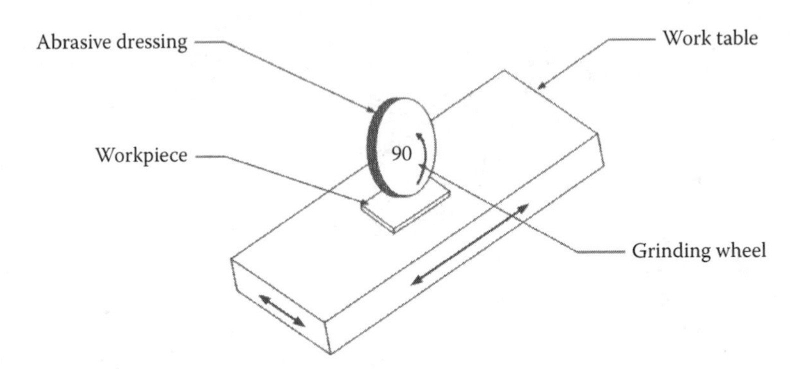

Figure 7.11 Grinding.

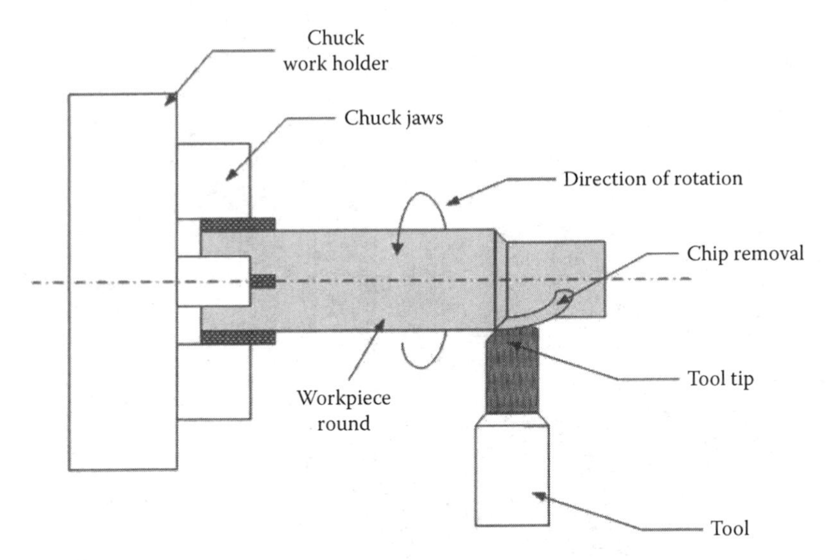

Figure 7.12 Turning.

promote effective cutting. Most notably, the cutting tool geometry is characterized by the rake angle, lead or main cutting-edge angle, nose radius, and edge radius which combined give the tool its signature. Common single-point operations include turning, boring, and facing. Turning is performed to make round parts, facing makes flat features, and boring turns nonstandard diameter, internal cylindrical surfaces. Multipoint (fixed geometry) operations include milling and drilling. Milling operations can be further categorized into face milling, peripheral or slab milling, and end milling. The face milling uses the face of the tool while slab milling uses the periphery of the cutter to effect the cutting action. These are typically applied to make flat features at a rate of material removal significantly higher than single-point operations such as shaping and planing. End milling cuts along the face as well as the periphery and is used for making slots and extensive contours. Drilling is used to make standard-sized holes with a cutter that has more than one active cutting edge. Reaming is a hole-finishing operation that follows drilling.

Machines and computer integration

Machine tools are used to carry out machining operations and are selected based on their ability to achieve a certain level of accuracy in size and geometry. Engine lathes are the most versatile machines and can perform several operations, including turning, facing, drilling, boring, threading, and chamfering. Milling machines can be of the horizontal or vertical type, depending on the configuration of the cutter relative to the workpiece.

Numerical control (NC) machines add programmable flexibility to standard lathes and mills. The tape media used in traditional NC machines have been replaced by inexpensive computers in the newer computer numerical control (CNC) machines of today. Machining centers with multitool magazines and Mill-Turn arrangements have advanced the state-of-the-art further. Industrial robots are primarily material handlers, but are also used for painting, welding, drilling, and so on. The industrial robots lend programmable flexibility similar to NC machines. Programming allows for the minimization of routine actions such as tool repositioning to beginning of cut in NCs and standard pick-and-place motions in industrial robots.

Flexible machine cells and flexible manufacturing systems configure multiple CNC machines and automated material handlers that include robots, conveyors, and automated guided vehicle systems within a single system to extend their flexibility while producing goods. Computer-aided design/computer-aided manufacturing is replaced by computer-integrated manufacturing (CIM) to improve integration at a systemic level. CIM systems and flexibility combined with rapid prototyping have led to agile manufacturing enterprises. These systems employ flexible fixturing and automated inspection. The most popular automated inspection systems employed in industry are the coordinate measuring machines (CMMs), machine vision systems, and optical projectors. CMMs are also used in the reverse engineering enterprise to reconstruct the design from the product. Parts are reverse engineered whenever the original drawings are no longer available or the original equipment manufacturer does not support the product anymore.

Joining processes

Joining processes (Figures 7.13 through 7.15) are employed in the manufacture of multipiece parts and assemblies. *Joining* processes include mechanical fastening (bolting and riveting), adhesive bonding, and welding processes. Welding processes use different sources of heat to cause

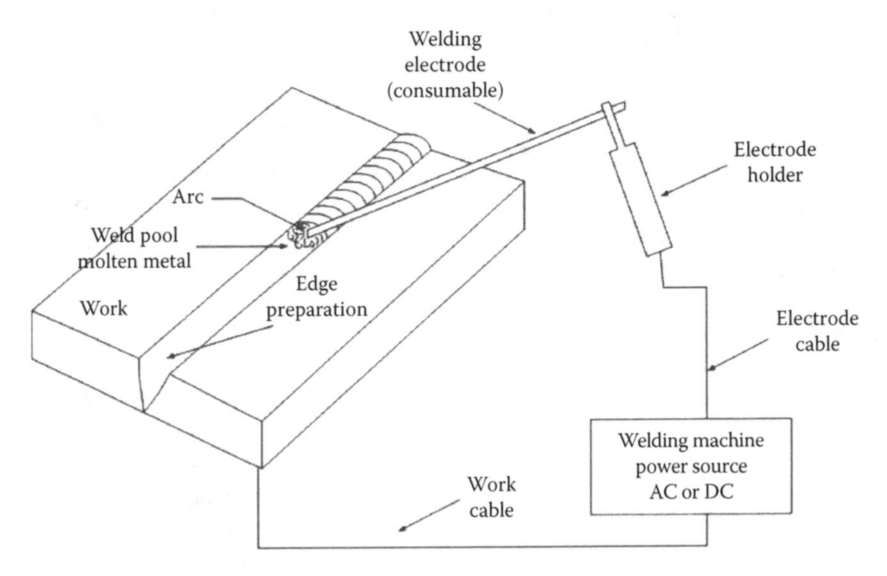

Figure 7.13 Welding.

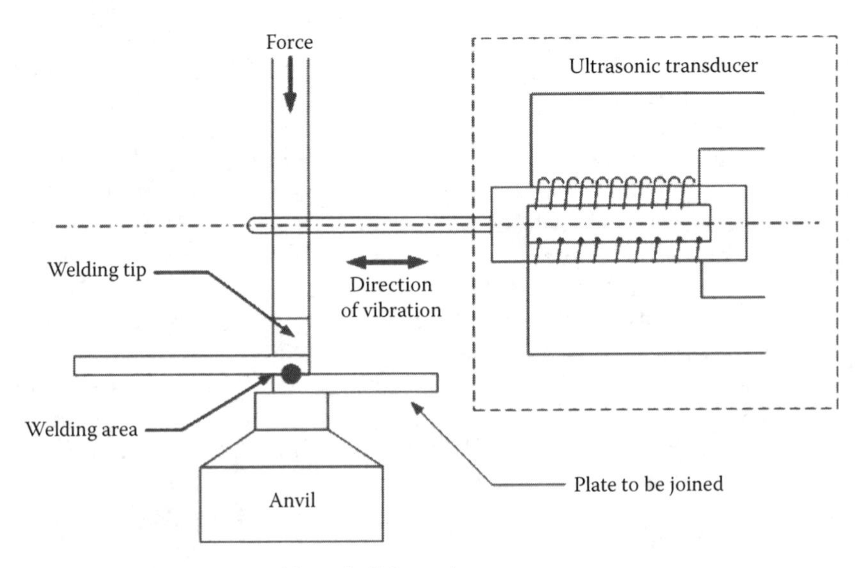

Figure 7.14 Ultrasonic welding (solid state).

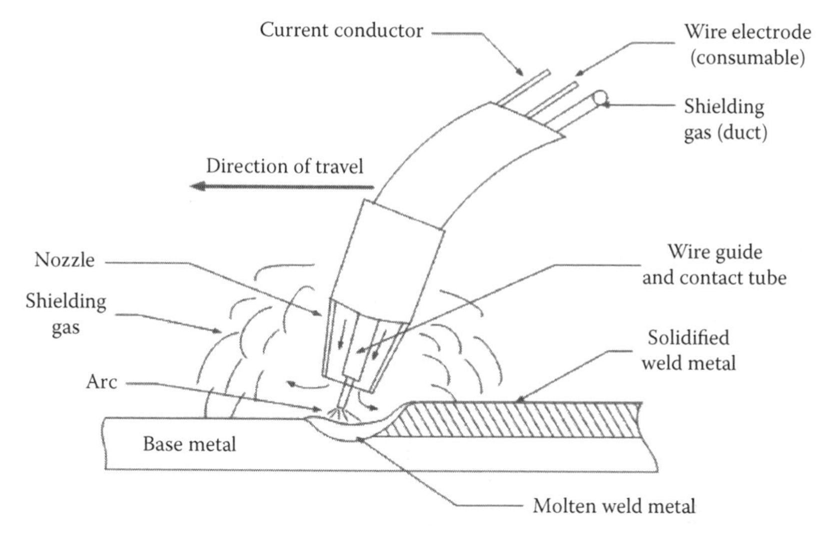

Figure 7.15 Gas metal arc welding.

localized melting of the metal to be joined or the melting of a filler to develop a joint between two metallic parts. Clean fraying surfaces are joined together through a butt-weld or a lap-weld, although other configurations are also possible. Two other joining processes are brazing and soldering, which differ from each other in the temperature applied.

The source of heating differs in different welding operations such as arc welding, gas welding, resistance, and solid-state welding. Arc welding strikes an arc between two electrodes to generate the requisite heat. One electrode is typically the plate to be joined. The other electrode could be consumable or nonconsumable. Stick welding is most common, and is also titled shielded metal arc welding (SMAW). Metal inert gas or gas metal arc welding, like SMAW, uses a consumable electrode. The electrode provides the filler and the inert gas provides an atmosphere such that contamination of the weld pool is prevented. A steady rate of flow of electrode is often made through automated means to maintain the arc gap thereby controlling the temperature of the arc. Gas tungsten arc welding or tungsten inert gas welding uses a non-consumable electrode and a separate filler must be supplied for welding.

In resistance welding, the resistance offered by the air gap between the fraying surfaces to the flow of electric current between two fixed electrodes is used to generate the heat required for welding. The focused heat can be used to make spot or seam welds. Gas welding typically employs acetylene and oxygen in various proportions to develop different temperatures to heat workpieces or fillers for welding, brazing, and soldering. If the acetylene is in excess, a reducing flame is obtained, whereas if oxygen is in excess, then an oxidizing flame is generated; if equal proportions are used, a neutral flame results. The oxidizing flames generate the highest temperatures. Other solid-state processes include thermit welding, ultrasonic welding, and friction welding.

Finishing processes

Finishing processes include polishing, shot-peening, sand blasting, cladding and electroplating, coating, and painting. Polishing may involve very minor material removal and is hence on occasion classified under machining operations. Shot and sand blasting are typically used to improve cleanliness and surface properties. Coatings are applied through chemical vapor deposition or physical vapor deposition to improve surface properties. For instance, in some cases, hard coatings are applied to softer and tougher substrates to improve wear resistance while retaining fracture resistance. The coatings are less than 10 microns thick in many cases. Cladding is done as in aluminum cladding on stainless steel to improve its heat conductivity.

Nontraditional manufacturing processes

Nontraditional processes include electrodischarge machining (EDM), electrochemical machining, chemical grinding, abrasive water-jet machining, and laser machining. The *electrodischarge machining*, termed earlier as spark erosion machining, uses a dielectric to control the machining rate. Softer electrodes may be employed to obtain complex-shaped (nonround) holes using EDM. Wire EDM is a variation of this process. *Abrasive jet machining* uses an abrasive slurry in a forced jet of water to machine hard and otherwise nonmachinable materials.

Even newer cutting-edge technologies have emerged, most notably, *nanotechnology* or *molecular manufacturing*. A nanometer is a billionth of a meter. The ability to modify and construct products at a molecular level by moving atoms and molecules to desired positions makes this very attractive. Single wall nanotubes are one of the biggest innovations that have been envisioned for building future transistors and sensors. Variants of the nano theme include nanometric fabrication including ultrahigh precision machining where very small depths of cut are consistently taken to fabricate components. Biotechnology, another emerging technology, also benefits significantly from nanotechnology and nanoscience.

Near net shape manufacturing is common in many applications, such as in the production of coins such as dimes and nickels. The idea is to complete much of the processing in a single step without requiring significant finishing. Injection molding, used for the fabrication of plastics; investment and impression-die casting; and precision forging are all considered to be near-net fabrication processes. Machining once considered wasteful and expensive has once again proven its immense worth in producing very tight tolerances and finishes. Powder processes are currently very important in the fabrication of very hard materials. *Powder metallurgy* is restricted to metals and involves a sequential application of compaction and sintering or isostatic compaction to shape objects. Other state-of-the-art technologies in rapid prototyping include stereolithography and 3D printing.

Conclusions

Manufacturing is the backbone and wealth of every developing country. Manufacturing processes are used to convert raw materials into useful products. Several resources such as humans, machines, tools, and tooling are employed in facilitating this conversion. In most cases, there is more than one way to manufacture a part to specifications. The decision is largely governed by variables including the value to be added, the costs of manufacturing and quality assurance, and desirable and achievable quality metrics. Casting, forming, machining, and welding are traditional processes employed extensively to this day. All the same, due to technological revolutions and the desire for more stringent specifications and better comfort, security, and safety, the terms *manufacturability* and *functionality* are continuously evolving. This has led to the development of several newer processes, such as nanotechnology, which could not have even been imagined 50 years ago. As our expectations continue to evolve, this trend of growth in processes and machines is expected to continue.

chapter eight

Manufacturing technology transfer strategies

> Technological progress is like an axe in the hands of a pathological criminal.
>
> **Albert Einstein**

The opening quote above reminds us that technology can easily be misused, if not properly controlled. Technology evolves for beneficial purposes, but its use can be misapplied, mistransferred, or misunderstood. Using appropriate transfer strategies can ensure that manufacturing technology can avoid the pathological criminal analogy conveyed in Albert Einstein's quote above.

Now that we have addressed the what, why, when, where, and when aspects of manufacturing technology in Chapters 1 through 7, we now address the *how* of transferring manufacturing technology from one endpoint to another. Technology transfer is not just about the hardware components of the technology. It can involve a combination of several components, including software (computer-based) and *skinware* (people-based). Thus, this chapter addresses the transfer of knowledge as well as the transfer of skills.

Why reinvent the wheel when it can be transferred and adopted from existing wheeled applications? The concepts of project management can be very helpful in planning for the adoption and implementation of new industrial technology. Due to its many interfaces, the area of industrial technology adoption and implementation is a prime candidate for the application of project planning and control techniques. Technology managers, engineers, and analysts should make an effort to take advantage of the effectiveness of project management tools. This applies the various project management techniques that have been discussed in Chapters 1 through 7 to the problem of industrial technology transfer. Project management approach is presented within the context of technology adoption and implementation for industrial development. Project management guidelines are presented for industrial technology management. The Triple C model of communication, cooperation, and coordination is applied as an effective tool for ensuring the acceptance of new technology. The importance of new technologies in improving product quality and operational productivity is also discussed. This chapter also outlines the strategies for project planning and control in complex technology-based operations.

Characteristics of technology transfer

To transfer technology, we must know what constitutes technology. A working definition of technology will enable us to determine how best to transfer it. A basic question that should be asked is *What is technology?*

Technology can be defined as follows:

Technology is a combination of physical and nonphysical processes that make use of the latest available knowledge to achieve business, service, or production goals.

Technology is a specialized body of knowledge that can be applied to achieve a mission or purpose. The knowledge concerned could be in the form of methods, processes, techniques, tools, machines, materials, and procedures. Technology design, development, and effective use are driven by effective utilization of human resources and effective management systems. Technological progress is the result obtained when the provision of technology is used in an effective and efficient manner to improve productivity, reduce waste, improve human satisfaction, and raise the quality of life.

Technology all by itself is useless. However, when the right technology is put to the right use, with effective supporting management system, it can be very effective in achieving industrialization goals. Technology implementation starts with an idea and ends with a productive industrial process. Technological progress is said to have occurred when the outputs of technology in the form of information, instrument, or knowledge that is used productively and effectively in industrial operations lead to a lowering of costs of production, better product quality, higher levels of output (from the same amount of inputs), and higher market share. The information and knowledge involved in technological progress includes those that improve the performance of management, labor, and the total resources expended for a given activity.

Technological progress plays a vital role in improving overall national productivity. Experience in developed countries such as in the United States shows that during the period 1870–1957, 90% of the rise in real output per man-hour can be attributed to technological progress. It is conceivable that a higher proportion of increases in per capita income is accounted for by technological change. Changes occur through improvements in the efficiency in the use of existing technology. That is, through learning and through the adaptation of other technologies, some of which may involve different collections of technological equipment. The challenge to developing countries is how to develop the infrastructure that promote, use, adapt, and advance technological knowledge.

Most of the developing nations today face serious challenges arising not only from the worldwide imbalance of dwindling revenue from industrial products and oil, but also from major changes in a world economy that is characterized by competition, imports, and exports not only of oil, but also of basic technology, weapon systems, and electronics. If technology utilization is not given the right attention in all sectors of the national economy, the much-desired industrial development cannot occur or cannot be sustained. The ability of a nation to compete in the world market will, consequently, be stymied.

The important characteristics or attributes of a new technology may include productivity improvement, improved quality, cost savings, flexibility, reliability, and safety. An integrated evaluation must be performed to ensure that a proposed technology is justified both economically and technically. The scope and goals of the proposed technology must be established right from the beginning of the project. Table 8.1 summarizes

Table 8.1 The *-ilities* of Manufacturing Technology

Characteristics	Definitions, questions, and implications
Adaptability	Can the technology be adapted to fit the needs of the organization? Can the organization adapt to the requirements of the technology?
Affordability	Can the organization afford the technology in terms of first-cost, installation cost, sustainment cost, and other incidentals?
Capability	What are the capabilities of the technology with respect to what the organization needs? Can the technology meet the current and emerging needs of the organization?
Compatibility	Is the technology compatible with existing software and hardware?
Configurability	Can the technology be configured for the existing physical infrastructure available within the organization?
Dependability	Is the technology dependable enough to produce the outputs expected?
Desirability	Is the particular technology desirable for the prevailing operating environment of the organization? Are there environmental issues and/or social concerns related the technology?
Expandability	Can the technology be expanded to fit the changing needs of the organization?

(Continued)

Table 8.1 (Continued) The *-ilities* of Manufacturing Technology

Characteristics	Definitions, questions, and implications
Flexibility	Does the technology have flexible characteristics to accomplish alternate production requirements?
Interchangeability	Can the technology be interchanged with currently available tools and equipment in the organization? In case of operational problems, can the technology be interchanged with something else?
Maintainability	Does the organization have the wherewithal to maintain the technology?
Manageability	Does the organization have adequate management infrastructure to acquire and use the technology?
Re-configurability	When operating conditions change or organizational infrastructure change, can the technology be re-configured to meet new needs?
Reliability	Is the technology reliable in terms of technical, physical, and/or scientific characteristics?
Stability	Is the technology mature and stable enough to warrant an investment within the current operating scenario?
Sustainability	Is the organization committed enough to sustain the technology for the long haul? Is the design of the technology sound and proven to be sustainable?
Volatility	Is the technology devoid of volatile developments? Is the source of the technology devoid of political upheavals and/or social unrests?

some of the common *-ilities* characteristics of technology transfer for a well-rounded assessment.

An assessment of a technology transfer opportunity will entail a comparison of departmental objectives with overall organizational goals in the following areas:

1. *Industrial marketing strategy*: This should identify the customers of the proposed technology. It should also address items such as market cost of proposed product, assessment of competition, and market share. Import and export considerations should be a key component of the marketing strategy.
2. *Industry growth and long-range expectations*: This should address short-range expectations, long-range expectations, future competitiveness, future capability, and prevailing size and strength of the industry that will use the proposed technology.
3. *National benefit*: Any prospective technology must be evaluated in terms of direct and indirect benefits to be generated by the technology. These may include product price versus value, increase

in international trade, improved standard of living, cleaner environment, safer work place, and higher productivity.

4. *Economic feasibility*: An analysis of how the technology will contribute to profitability should consider past performance of the technology, incremental benefits of the new technology versus conventional technology, and value added by the new technology.
5. *Capital investment*: Comprehensive economic analysis should play a significant role in the technology assessment process. This may cover an evaluation of fixed and sunk costs, cost of obsolescence, maintenance requirements, recurring costs, installation cost, space requirement cost, capital substitution options, return on investment, tax implications, cost of capital, and other concurrent projects.
6. *Resource requirements*: The utilization of resources (human resources and equipment) in the pre-technology and post-technology phases of industrialization should be assessed. This may be based on material input–output flows, high value of equipment versus productivity improvement, required inputs for the technology, expected output of the technology, and utilization of technical and nontechnical personnel.
7. *Technology stability*: Uncertainty is a reality in technology adoption efforts. Uncertainty will need to be assessed for the initial investment, return on investment, payback period, public reactions, environmental impact, and volatility of the technology.
8. *National productivity improvement*: An analysis of how the technology may contribute to national productivity may be verified by studying industrial throughput, efficiency of production processes, utilization of raw materials, equipment maintenance, absenteeism, learning rate, and design-to-production cycle.

Emergence of new technology

New industrial and service technologies have been gaining more attention in recent years. This is due to the high rate at which new productivity improvement technologies are being developed. The fast pace of new technologies has created difficult implementation and management problems for many organizations. New technology can be successfully implemented only if it is viewed as a system whose various components must be evaluated within an integrated managerial framework. Such a framework is provided by a project management approach. A multitude of new technologies has emerged in recent years. It is important to consider the peculiar characteristics of a new technology before establishing adoption and implementation strategies. The justification for the adoption of a new technology is usually a combination of several factors rather than a single characteristic of the technology. The potential of a specific technology to contribute to industrial development goals must be carefully assessed. The technology assessment process should explicitly address the following questions:

- What is expected from the new technology?
- Where and when will the new technology be used?
- How is the new technology similar to or different from existing technologies?
- What is the availability of technical personnel to support the new technology?
- What administrative support is needed for the new technology?
- Who will use the new technology?
- How will the new technology be used?
- Why is the technology needed?

The development, transfer, adoption, utilization, and management of technology is a problem that is faced in one form or another by business, industry, and government establishments. Some of the specific problems in technology transfer and management include the following:

- Controlling technological change
- Integrating technology objectives
- Shortening the technology transfer time
- Identifying a suitable target for technology transfer
- Coordinating the research and implementation interface
- Formal assessment of current and proposed technologies
- Developing accurate performance measures for technology
- Determining the scope or boundary of technology transfer
- Managing the process of entering or exiting a technology
- Understanding the specific capability of a chosen technology
- Estimating the risk and capital requirements of a technology

Integrated managerial efforts should be directed at the solution of the problems stated above. A managerial revolution is needed in order to cope with the ongoing technological revolution. The revolution can be initiated by modernizing the long-standing and obsolete management culture relating to technology transfer. Some of the managerial functions that will need to be addressed when developing a technology transfer strategy include the following:

1. Development of a technology transfer plan.
2. Assessment of technological risk.
3. Assignment/reassignment of personnel to implement the technology transfer.
4. Establishment of a transfer manager and a technology transfer office. In many cases, transfer failures occur because no individual has been given the responsibility to ensure the success of technology transfer.

5. Identification and allocation of the resources required for technology transfer.
6. Setting of guidelines for technology transfer. For example,
   a. Specification of phases (development, testing, transfer, etc.)
   b. Specification of requirements for inter-phase coordination
   c. Identification of training requirements
   d. Establishment and implementation of performance measurement
7. Identify key factors (both qualitative and quantitative) associated with technology transfer and management.
8. Investigate how the factors interact and develop the hierarchy of importance for the factors.
9. Formulate a loop system model that considers the forward and backward chains of actions needed to effectively transfer and manage a given technology.
10. Track the outcome of the technology transfer.

Technological developments in many industries appear in scattered, narrow, and isolated areas within a few selected fields. This makes technology efforts to be rarely coordinated, thereby, hampering the benefits of technology. The optimization of technology utilization is, thus, very difficult. To overcome this problem and establish the basis for effective technology transfer and management, an integrated approach must be followed. An integrated approach will be applicable to technology transfer between any two organizations whether public or private.

Some nations concentrate on the acquisition of bigger, better, and faster technology. But little attention is given to how to manage and coordinate the operations of the technology once it arrives. When technology fails, it is not necessarily because the technology is deficient. Rather, it is often the communication, cooperation, and coordination functions of technology management that are deficient. Technology encompasses factors and attributes beyond mere hardware, software, and *skinware*, which refers to people issues affecting the utilization of technology. This may involve socioeconomic and sociocultural issues of using certain technologies. Consequently, technology transfer involves more than the physical transfer of hardware and software. Several flaws exist in the common practices of technology transfer and management. These flaws include in the following:

- *Poor fit*: This relates to an inadequate assessment of the need of the organization receiving the technology. The target of the transfer may not have the capability to properly absorb the technology.
- *Premature transfer of technology*: This is particularly acute for emerging technologies that are prone to frequent developmental changes.

- *Lack of focus*: In the attempt to get a bigger share of the market or gain early lead in the technological race, organizations frequently force technology in many incompatible directions.
- *Intractable implementation problems*: Once a new technology is in place, it may be difficult to locate sources of problems that have their roots in the technology transfer phase itself.
- *Lack of transfer precedents*: Very few precedents are available on the management of brand new technology. Managers are, thus, often unprepared for their new technology management responsibilities.
- *Stuck on technology*: Unworkable technologies sometimes continue to be recycled needlessly in the attempt to find the *right* usage.
- *Lack of foresight*: Due to the nonexistence of a technology transfer model, managers may not have a basis against which they can evaluate future expectations.
- *Insensitivity to external events*: Some external events that may affect the success of technology transfer may include trade barriers, taxes, and political changes.
- *Improper allocation of resources*: There is usually not enough resources available to allocate to technology alternatives. Thus, a technology transfer priority must be developed.

The following steps provide a specific guideline for pursuing the implementation of manufacturing technology transfer:

1. Find a suitable application
2. Commit to an appropriate technology
3. Perform economic justification
4. Secure management support for the chosen technology
5. Design the technology implementation to be compatible with existing operations
6. Formulate project management approach to be used
7. Prepare the receiving organization for the technology change
8. Install the technology
9. Maintain the technology
10. Periodically review the performance of the technology based on prevailing goals

Technology transfer modes

The transfer of technology can be achieved in various forms. Project management provides an effective means of ensuring proper transfer of technology. Three technology transfer modes are presented here to illustrate basic strategies for getting one technological product from one point (technology source) to another point (technology sink). A conceptual

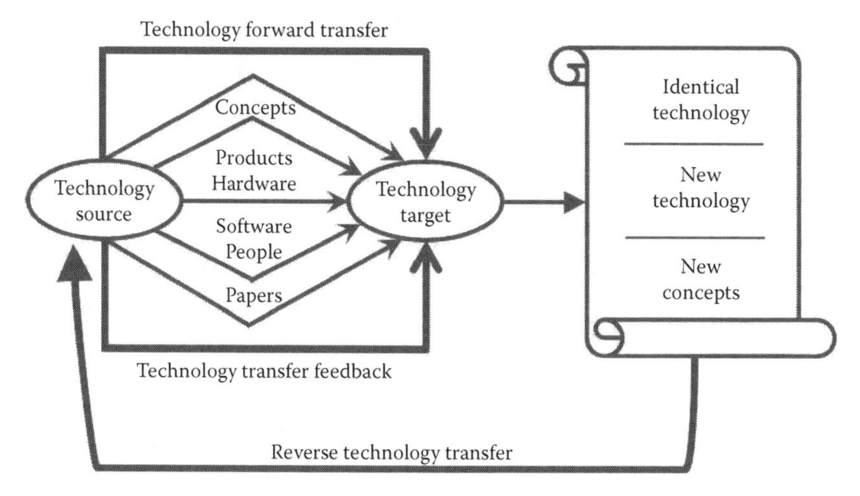

Figure 8.1 Technology transfer modes.

integrated model of the interaction between the technology source and sink is presented in Figure 8.1.

The university–industry interaction model presented in this book can be used as an effective mechanism for facilitating technology transfer. Industrial technology application centers may be established to serve as a unified point for linking technology sources with interested targets. The center will facilitate interactions between business establishments, academic institutions, and government agencies to identify important technology needs. With reference to Figure 8.1, technology can be transferred in one or a combination of the following strategies:

1. *Transfer of complete technological products*: In this case, a fully developed product is transferred from a source to a target. Very little product development effort is carried out at the receiving point. However, information about the operations of the product is fed back to the source so that necessary product enhancements can be pursued. So, the technology recipient generates product information that facilitates further improvement at the technology source. This is the easiest mode of technology transfer and the most tempting. Developing nations are particularly prone to this type of transfer. Care must be exercised to ensure that this type of technology transfer does not degenerate into *machine transfer*. It should be recognized that machines alone do not constitute technology.
2. *Transfer of technology procedures and guidelines*: In this technology transfer mode, procedures (e.g., blueprints) and guidelines are transferred from a source to a target. The technology blueprints

are implemented locally to generate the desired services and products. The use of local raw materials and manpower is encouraged for the local production. Under this mode, the implementation of the transferred technology procedures can generate new operating procedures that can be fed back to enhance the original technology. With this symbiotic arrangement, a loop system is created whereby both the transferring and the receiving organizations derive useful benefits.

3. *Transfer of technology concepts, theories, and ideas*: This strategy involves the transfer of the basic concepts, theories, and ideas behind a given technology. The transferred elements can then be enhanced, modified, or customized within local constraints to generate new technological products. The local modifications and enhancements have the potential to generate an identical technology, a new related technology, or a new set of technology concepts, theories, and ideas. These derived products may then be transferred back to the original technology source as new technological enhancements. Figure 8.2 presents a specific cycle for local adaptation and modification of technology. An academic institution is a good potential source for the transfer of technology concepts, theories, and ideas.

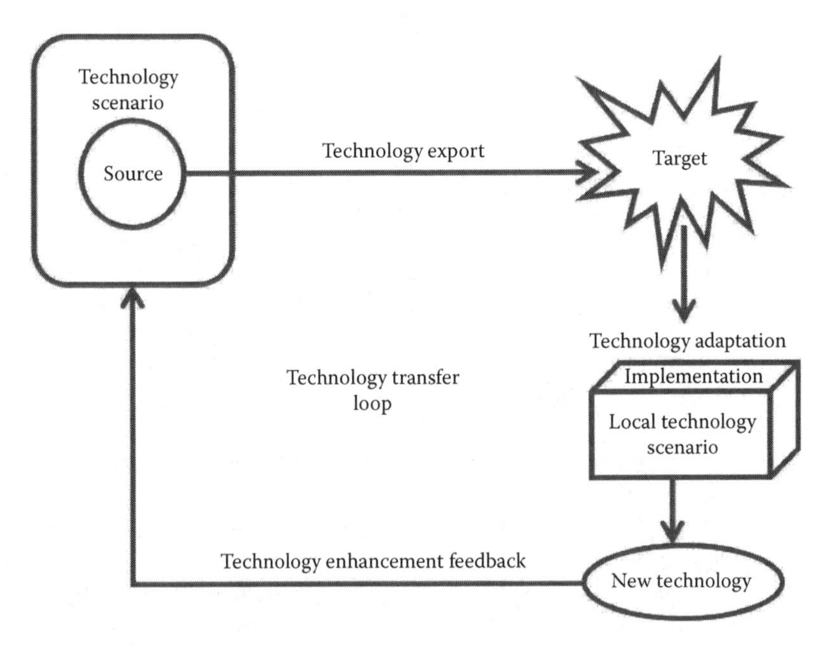

Figure 8.2 Local adaptation and enhancement of technology.

It is very important to determine the mode in which technology will be transferred for manufacturing purposes. There must be a concerted effort by people to make the transferred technology work within local infrastructure and constraints. Local innovation, patriotism, dedication, and willingness to adapt technology will be required to make technology transfer successful. It will be difficult for a nation to achieve industrial development through total dependence on transplanted technology. Local adaptation will always be necessary.

Technology changeover strategies

Any development project will require changing from one form of technology to another. The implementation of a new technology to replace an existing (or a nonexistent) technology can be approached through one of several options. Some options are more suitable than others for certain types of technologies. The most commonly used technology changeover strategies include the following:

- *Parallel changeover*: In this case, the existing technology and the new technology operate concurrently until there is confidence that the new technology is satisfactory.
- *Direct changeover*: In this approach, the old technology is removed totally and the new technology takes over. This method is recommended only when there is no existing technology or when both technologies cannot be kept operational due to incompatibility or cost considerations.
- *Phased changeover*: In this incremental changeover method, modules of the new technology are gradually introduced one at a time using either direct or parallel changeover.
- *Pilot changeover*: In this case, the new technology is fully implemented on a pilot basis in a selected department within the organization.

Post-implementation evaluation

The new technology should be evaluated only after it has reached a steady-state performance level. This helps to avoid the bias that may be present at the transient stage due to personnel anxiety, lack of experience, or resistance to change. The system should be evaluated for the following aspects:

- Sensitivity to data errors
- Quality and productivity
- Utilization level
- Response time
- Effectiveness

Technology systems integration

With the increasing shortages of resources, more emphasis should be placed on the sharing of resources. Technology resource sharing can involve physical equipment, facilities, technical information, ideas, and related items. The integration of technologies facilitates the sharing of resources. Technology integration is a major effort in technology adoption and implementation. Technology integration is required for proper product coordination. Integration facilitates the coordination of diverse technical and managerial efforts to enhance organizational functions, reduce cost, improve productivity, and increase the utilization of resources. Technology integration ensures that all performance goals are satisfied with a minimum of expenditure of time and resources. It may require the adjustment of functions to permit sharing of resources, development of new policies to accommodate product integration, or realignment of managerial responsibilities. It can affect both hardware and software components of an organization. Important factors in technology integration include the following:

- Unique characteristics of each component in the integrated technologies
- Relative priorities of each component in the integrated technologies
- How the components complement one another
- Physical and data interfaces between the components
- Internal and external factors that may influence the integrated technologies
- How the performance of the integrated system will be measured

Role of government in technology transfer

The malignant policies and operating characteristics of some of the governments in underdeveloped countries have contributed to stunted growth of technology in those parts of the world. The governments in most developing countries control the industrial and public sectors of the economy. Either people work for the government or serve as agents or contractors for the government. The few industrial firms that are privately owned depend on government contracts to survive. Consequently, the nature of the government can directly determine the nature of industrial technological progress.

The operating characteristics of most of the governments perpetuate inefficiency, corruption, and bureaucratic bungles. This has led to a decline in labor and capital productivity in the industrial sectors. Using the Pareto distribution, it can be estimated that in most government-operated companies, there are eight administrative workers for every two production workers. This creates a nonproductive environment that is skewed toward hyper-bureaucracy. The government of a nation pursuing industrial development must formulate and maintain an economic stabilization policy. The objective should be to minimize the sacrifice of economic growth in the short run while maximizing long-term economic growth. To support industrial technology transfer efforts, it is essential that a conducive national policy be developed.

More emphasis should be placed on industry diversification, training of the work force, supporting financial structure for emerging firms, and implementing policies that encourage productivity in a competitive economic environment. Appropriate foreign exchange allocation, tax exemptions, bank loans for emerging businesses, and government-guaranteed low interest loans for potential industrial entrepreneurs are some of the favorable policies to spur growth and development of the industrial sector.

Improper trade and domestic policies have adversely affected industrialization in many countries. Excessive regulations that cause bottlenecks in industrial enterprises are not uncommon. The regulations can take the form of licensing, safety requirements, manufacturing value-added quota requirements, capital contribution by multinational firms, and high domestic production protection. Although regulations are needed for industrial operations, excessive controls lead to low returns from the industrial sectors. For example, stringent regulations on foreign exchange allocation and control have led to the closure of industrial plants in some countries. The firms that cannot acquire essential raw materials, commodities, tools, equipment, and new technology from abroad due to foreign exchange restrictions are forced to close and lay off workers.

Price controls for commodities are used very often by developing countries especially when inflation rates for essential items are high. The disadvantages involved in price control of industrial goods include restrictions of the free competitive power of available goods in relation to demand and supply, encouragement of inefficiency, promotion of dual markets, distortion of cost relationships, and increase in administrative costs involved in producing goods and services.

U.S. templates for technology transfer

One way that a government can help facilitate industrial technology transfer involves the establishment of technology transfer centers within appropriate government agencies. A good example of this approach can be seen in the government-sponsored technology transfer program by the U.S. National Aeronautics and Space Administration (NASA). In the Space Act of 1958, the U.S. Congress charged NASA with a responsibility to provide for the widest practical and appropriate dissemination of information concerning its activities and the results achieved from those activities. With this technology transfer responsibility, technology developed in the United State's space program is available for use by the nation's business and industry.

In order to accomplish technology transfer to industry, NASA established a Technology Utilization Program (TUP) in 1962. The TUP uses several avenues to disseminate information on NASA technology. The avenues include the following:

- Complete, clear, and practical documentation is required for new technology developed by NASA and its contractors. These are available to industry through several publications produced by NASA. An example is a monthly, Tech Briefs, which outlines technology innovations. This is a source of prompt technology information for industry.
- Industrial Application Centers (IAC) were developed to serve as repositories for vast computerized data on technical knowledge. The IACs are located at academic institutions around the country. All the centers have access to a large data base containing millions of NASA documents. With this data base, industry can have access to the latest technological information quickly. The funding for the centers are obtained through joint contributions from several sources including NASA, the sponsoring institutions, and state government subsidies. Thus, the centers can provide their services at very reasonable rates.
- NASA operates a Computer Software Management and Information Center (COSMIC) to disseminate computer programs developed through NASA projects. COSMIC, which is located at a university, has a library of thousands of computer programs. The center publishes an annual index of available software.

In addition to the specific mechanisms discussed above, NASA undertakes application engineering projects. Through these projects, NASA collaborates with industry to modify aerospace technology for use in industrial applications. To manage the application projects, NASA established a technology application team, consisting of scientists and engineers from several disciplines. The team interacts with NASA field centers, industry, universities, and government agencies. The major mission of the team interactions is to define important technology needs and identify possible solutions within NASA. NASA applications engineering projects are usually developed in a five-phase approach with go or no-go decisions made by NASA and industry at the completion of each phase. The five phases are outlined below:

1. NASA and the technology application team meet with industry associations, manufacturers, university researchers, and public sector agencies to identify important technology problems that might be solved by aerospace technology.
2. After a problem is selected, it is documented and distributed to the technology utilization officer at each of NASA's field centers.

The officer in turn distributes the description of the problem to the appropriate scientists and engineers at the center. Potential solutions are forwarded to the team for review. The solutions are then screened by the problem originator to assess the chances for technical and commercial success.

3. The development of partnerships and a project plan to pursue the implementation of the proposed solution. NASA joins forces with private companies and other organizations to develop an applications engineering project. Industry participation is encouraged through a variety of mechanisms such as simple letters of agreement or joint endeavor contracts. The financial and technical responsibilities of each organization are specified and agreed upon.
4. At this point, NASA's primary role is to provide technical assistance to facilitate utilization of the technology. The costs for these projects are usually shared by NASA and the participating companies. The proprietary information provided by the companies and their rights to new discoveries are protected by NASA.
5. The final phase involves the commercialization of the product. With the success of commercialization, the project would have widespread impact. Usually, the final product development, field testing, and marketing are managed by private companies without further involvement from NASA.

Through this well-coordinated government-sponsored technology transfer program, NASA has made significant contributions to the U.S. industry. The results of NASA's technology transfer abound in numerous consumer products either in subtle forms or in clearly identifiable forms. Food preservation techniques constitute one area of NASA's technology transfer that has had a significant positive impact on the society. Although the specific organization and operation of the NASA technology transfer programs have changed in name or in deed over the years, the basic descriptions outlined above remain a viable template for how to facilitate manufacturing technology transfer. Other nations can learn from NASA's technology transfer approach. In a similar government-backed strategy, the U.S. Air Force Research Lab also has very structured programs for transferring nonclassified technology to the industrial sector.

The major problem in developing nations is not the lack of good examples to follow. Rather, the problem involves not being able to successfully manage and sustain a program that has proven successful in other nations. It is believed that a project management approach can help in facilitating success with manufacturing technology transfer efforts.

Pathway to national strategy

Most of the developing nations depend on technologies transferred from developed nations to support their industrial base. This is partly due to a lack of local research and development programs, development funds, and workforce needed to support such activity. Advanced technology is desired by most industries in developing countries because of its potential to increase output. The adaptability of advanced technology to industries in a developing country is a complex and difficult task. Evidence in most manufacturing firms that operate in developing countries reveals that advanced technology can lead to machine downtime because the local plants do not have the maintenance and repair facilities to support the use of advanced technology.

In some situations, most firms cannot afford the high cost of maintenance associated with the use of foreign technology. One way to solve the transfer of technology problem is by establishing local design centers for developing nation's industrial sectors, such centers will design and adapt technology for local usage. In addition, such centers will also work on adapting full assembled machinery from developed countries. However, the fertile ground for the introduction of appropriate technology is where people are already organized under a good system of government, production, marketing, and continuing improvement in standard of living. Developing countries must place more emphasis on the production of useful, consumable goods and services. One useful strategy to ensure a successful transfer of technology is by providing training services that will ensure proper repair and maintenance of technology hardware. It is important that a nation trying to transfer technology should have access to a broad-based body of technical information and experience. A plan of technical information sharing between suppliers and users must be assured. The transfer of technology also requires a reliable liaison between the people who develop the ideas, their agents, and the people who originate the concepts. Technology transfer is only complete when the technology becomes generally accepted in the work place. Local efforts are needed in tailoring technological solutions to local problems. Technicians and engineers must be trained to assume the role of technology custodians so that implementation and maintenance problems are minimized. A strategy for minimizing the technology transfer disconnection is to set up central repair shops dedicated to making spare parts and repairing equipment on a timely basis to reduce industrial machine downtime. If the utilization level of equipment is increased, there will be an increase in the productive capacity of the manufacturer. Improving maintenance and repair centers in developing countries will provide an effective way of assisting emerging firms in developing countries where dependence on transferred technology is prevalent. There should also be a strategy to develop appropriate local technology to support the goals of industrialization. This is important because fully transferred technology may not be fully suitable or compatible with local product specifications. For example, many nations have experienced the failure of transferred food processing technology because the technology was not responsive to the local diets, ingredients, and food preparation practices. One way to accomplish the development of local technology is to encourage joint research efforts between academic institutions and industrial firms. Chapter 11 explicitly address university–industry collaborations. The design centers suggested earlier can help in this process. Chapter 13 presents a case example of the Industrial Development Center in Nigeria. In addition to developing new local technologies, existing technologies should be calibrated for local usage and the higher production level required for industrialization.

The government of developing nations must assume leadership roles in encouraging research and development activities, awarding research grants to universities and private organizations geared toward seeking better ways for developing and adapting technologies for local usage. Effective innovations and productivity improvement cannot happen without adequate public and private sector policies. A nation that does not have an effective policy for productivity management and technology advancement will always find itself in a cycle of unstable economy and business crisis. Increases in real product capital, income level, and quality of life are desirable goals that are achievable through effective policies that are executed properly. The following recommendations are offered to encourage industrial growth and technological progress:

1. Encourage free enterprise system that believes in and practice fair competition. Discourage protectionism and remove barriers to allow free trade.
2. Avoid nationalization of assets of companies jointly developed by citizens of developing countries and multinationals. Encourage joint industrial ventures among nations.
3. Both public and private sectors of the economy should encourage and invest in improving national education standards for citizen at various levels.
4. Refrain from dependence on borrowed money and subsidy programs. Create productive enterprises locally that provide essential commodities for local consumptions and exports.
5. Both public and private sectors should invest more on systems and programs, research and development that generate new breakthrough in technology, and methods for producing food rather than war instruments.

6. The public sector should establish science and technology centers to foster the development of new local technology, productivity management techniques, and production methodologies.
7. Encourage strong partnership between government, industry, and academic communities in formulating and executing national development programs.
8. Governments and financial institutions should provide low interest loans to entrepreneurs willing to take risk in producing essential goods and services through small-scale industries.
9. Implement a tax structure that is equitable and one that provides incentives for individuals and businesses that are working to expand employment opportunities and increase the final output of the national economy.
10. Refrain from government control of productive enterprises. Such controls only create grounds for fraud and corruption. Excessive regulations should be discouraged.
11. Periodically assess the ratio of administrative workers to production workers, administrative workers to service workers, in both private and public sectors. Implement actions to reduce excessive administrative procedures and bureaucratic bottlenecks that impede productivity and technological progress.
12. Encourage organizations and firms to develop and implement strategies, methods, and techniques in a framework of competitive and long-term performance.
13. Trade policy laws and regulations should be developed and enforced in a framework that recognizes fair competition in a global economy.
14. Create a national productivity, science and technology council to facilitate the implementation of good programs, enhance cooperation between private and public sectors of the economy, redirect the economy toward growth strategies, and encourage education and training of the work force.
15. Implement actions that insure stable fiscal, monetary, and income policies. Refrain from wage and price control by political means. Let the elements of the free enterprise system control inflation rate, wages, and income distribution.
16. Encourage morale standards that take pride in excellence, work ethics, and value system that encourage pride in consumer products produced locally.
17. Encourage individuals and business to protect full employment programs, maintain income levels by investing in local ventures rather than exporting capital abroad.
18. Both the public and private sectors of the economy should encourage and invest in re-training of the workforce as new technology and techniques are introduced for productive activities.

19. Make use of the expertise of nations that are professionally based abroad. This is an excellent source of expertise for local technology development.
20. Arrange for annual conferences, seminars, and workshops to exchange ideas between researchers, entrepreneurs, practitioners, and managers with the focus on the processes required for industrial development.

Using PICK chart for technology selection

The question of which technology is appropriate to transfer in or transfer out is relevant for technology transfer considerations. While several methods of technology selection are available, this book recommends methods that combine qualitative and quantitative factors. The analytical hierarchy process (AHP) is one such method. Another useful, but less publicized is the PICK chart. The PICK chart was originally developed by Lockheed Martin to identify and prioritize improvement opportunities in the company's process improvement applications. The technique is just one of the several decision tools available in process improvement endeavors. It is a very effective technology selection tool used to categorize ideas and opportunities. The purpose is to qualitatively help identify the most useful ideas. A 2 × 2 grid is normally drawn on a white board or large flip chart. Ideas that were written on sticky notes by team members are placed on the grid based on a group assessment of the payoff relative the level of difficulty. The PICK acronym comes from the labels for each of the quadrants of the grid: *P*ossible (easy, low payoff), *I*mplement (easy, high payoff), *C*hallenge (hard, high payoff), and *K*ill (hard, low payoff). The PICK chart quadrants are summarized as follows:

*P*ossible (easy, low payoff)	Third quadrant
*I*mplement (easy, high payoff)	Second quadrant
*C*hallenge (hard, high payoff)	First quadrant
*K*ill (hard, low payoff).	Fourth quadrant

The primary purpose is to help identify the most useful ideas, especially those that can be accomplished immediately with little difficulty. These are called *Just-Do-Its*. The general layout of the PICK chart grid is shown in Figure 8.3. The PICK process is normally done subjectively by a team of decision makers under a group decision process. This can lead to bias and protracted debate of where each item belongs. It is desired to improve the efficacy of the process by introducing some quantitative analysis. Badiru and Thomas (2013) present a methodology to achieve a quantification of the PICK selection process. The PICK chart is often

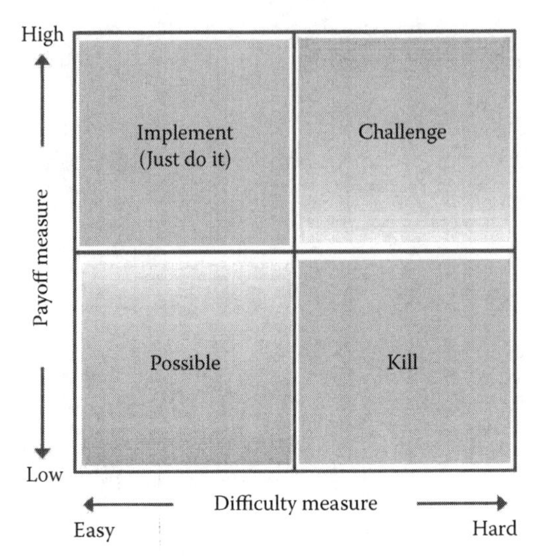

Figure 8.3 Basic layout of the PICK chart.

criticized for its subjective rankings and lack of quantitative analysis. The approach presented by Badiru and Thomas (2013) alleviates such concerns by normalizing and quantifying the process of integrating the subjective rakings by those involved in the group PICK process. Human decision is inherently subjective. All we can do is to develop techniques to mollify the subjective inputs rather than compounding them with subjective summarization.

PICK chart quantification methodology

The placement of items into one of the four categories in a PICK chart is done through expert ratings, which are often subjective and non-quantitative. In order to put some quantitative basis to the PICK chart analysis, Badiru and Thomas (2013) present the methodology of dual numeric scaling on the impact and difficulty axes. Suppose each technology is ranked on a scale of one to ten and plotted accordingly on the PICK chart. Then, each project can be evaluated on a binomial pairing of the respective rating on each scale. Note that a high rating along the x axis is desirable, while a high rating along the y axis is not desirable. Thus, a composite rating involving x and y must account for the adverse effect of high values of y. A simple approach is to define $y' = (11-y)$, which is then used in the composite evaluation. If there are more factors involved in the overall project selection scenario, the other factors can take on their own lettered labeling (e.g., a, b, c, z, etc.). Then, each project will have an n-factor assessment vector. In its simplest form, this approach will generate a rating such as the following:

$$\text{PICK}_{R,i}(x, y') = x + y'$$

where:

$\text{PICK}_{R,i}(x,y)$ = PICK rating of project i ($i = 1, 2, 3, \ldots, n$)
n is the number of project under consideration
x is the rating along the impact axis ($1 \leq x \leq 10$)
y is the rating along the difficulty axis ($1 \leq y \leq 10$)
$y' = (11-y)$

If $x + y'$ is the evaluative basis, then each technology's composite rating will range from 2 to 20, 2 being the minimum and 20 being the maximum possible. If $(x)(y)$ is the evaluative basis, then each project's composite rating will range from 1 to 100. In general, any desired functional form may be adopted for the composite evaluation. Another possible functional form is

$$\begin{aligned}\text{PICK}_{R,i}(x,y'') &= f(x,y'') \\ &= (x+y'')^2\end{aligned}$$

where:

y'' is defined as needed to account for the converse impact of the axes of difficulty

The above methodology provides a quantitative measure for translating the entries in a conventional PICK chart into an analytical technique to rank the technology alternatives, thereby reducing the level of subjectivity in the final decision. The methodology can be extended to cover cases where a technology has the potential to create negative impacts, which may impede organizational advancement.

The quantification approach facilitates a more rigorous analytical technique compared to traditional subjective approaches. One concern is that although quantifying the placement of alternatives on the PICK chart may improve the granularity of relative locations on the chart, it still does not eliminate the subjectivity of how the alternatives are assigned to quadrants in the first place. This is a recognized feature of many decision tools. This can be mitigated by the use of additional techniques that aid decision makers to refine their choices. The AHP could be useful for this purpose. Quantifying subjectivity is a continuing challenge in decision analysis. The PICK chart quantification methodology offers an improvement over the conventional approach.

Although the PICK chart has been used extensively in industry, there are few published examples in the open literature. The quantification approach presented by Badiru and Thomas (2013) may expand interest and applications of the PICK chart among technology researchers and practitioners. The steps for implementing a PICK chart are summarized as follows:

Step 1: On a chart, place the subject question. The question needs to be asked and answered by the team at different stages to be sure that the data collected are relevant.

Step 2: Put each component of the data on a different note like a post-it or small cards. These notes should be arranged on the left side of the chart.

Step 3: Each team member must read all notes individually and consider its importance. The team member should decide whether the element should or should not remain a fraction of the significant sample. The notes are then removed and moved to the other side of the chart. Now, the data are condensed enough to be processed for a particular purpose by means of tools that allow groups to reach a consensus on priorities of subjective and qualitative data.

Step 4: Apply the quantification methodology presented above to normalize the qualitative inputs of the team.

DEJI model for technology integration

Technology is at the intersection of efficiency, effectiveness, and productivity. Efficiency provides the framework for quality in terms of resources and inputs required to achieve the desired level of quality. Effectiveness comes into play with respect to the application of product quality to meet specific needs and requirements of an organization. Productivity is an essential factor in the pursuit of quality as it relates to the throughput of a production system. To achieve the desired levels of quality, efficiency, effectiveness, and productivity, a new technology integration framework must be adopted. This section presents a technology integration model for design, evaluation, justification, and integration (DEJI) based on the product development application presented by Badiru (2012). The model is relevant for research and development efforts in industrial development and technology applications. The DEJI model encourages the practice of building quality into a product right from the beginning so that the product or technology integration stage can be more successful. The essence of the model is summarized in Table 8.2.

Table 8.2 DEJI Model for Technology Integration

DEJI model	Characteristics	Tools and techniques
Design	Define goals Set performance metrics Identify milestones	Parametric assessment Project state transition Value stream analysis
Evaluate	Measure parameters Assess attributes Benchmark results	Pareto distribution Life cycle analysis Risk assessment
Justify	Assess economics Assess technical output Align with goals	Benefit–cost ratio Payback period Present value
Integrate	Embed in normal operation Verify symbiosis Leverage synergy	SMART concept Process improvement Quality control

Design for technology implementation

The design of quality in product development should be structured to follow point-to-point transformations. A good technique to accomplish this is the use of state-space transformation, with which we can track the evolution of a product from the concept stage to a final product stage. For the purpose of product quality design, the following definitions are applicable:

- *Product state*: A state is a set of conditions that describe the product at a specified point in time. The *state* of a product refers to a performance characteristic of the product that relates input to output such that a knowledge of the input function over time and the state of the product at time $t = t_0$ determines the expected output for $t \geq t_0$. This is particularly important for assessing where the product stands in the context of new technological developments and the prevailing operating environment.
- *Product state space*: A product *state-space* is the set of all possible states of the product life cycle. State-space representation can solve product design problems by moving from an initial state to another state, and eventually to the desired end-goal state. The movement from state to state is achieved by means of actions. A goal is a description of an intended state that has not yet been achieved. The process of solving a product problem involves finding a sequence of actions that represents a solution path from the initial state to the goal state. A state-space model consists of state variables that describe the prevailing condition of the product. The state variables are related to inputs by mathematical relationships. Examples of potential product state variables include schedule, output quality, cost, due date,

resource, resource utilization, operational efficiency, productivity throughput, and technology alignment. For a product described by a system of components, the state-space representation can follow the quantitative metric below:

$$Z = f(z, x); \;\; Y = g(z, x)$$

where:

f and g are vector-valued functions

The variable Y is the output vector while the variable x denotes the inputs

The state vector Z is an intermediate vector relating x to y

In generic terms, a product is transformed from one state to another by a driving function that produces a transitional relationship given by:

$$S_s = f\left(x \mid S_p\right) + e$$

where:

S_s is the subsequent state

x is the state variable

S_p is the the preceding state

e is the error component

The function f is composed of a given action (or a set of actions) applied to the product. Each intermediate state may represent a significant milestone in the project. Thus, a descriptive state-space model facilitates an analysis of what actions to apply in order to achieve the next desired product state. The state-space representation can be expanded to cover several components within the technology integration framework. Hierarchical linking of product elements provides an expanded transformation structure. The product state can be expanded in accordance with implicit requirements. These requirements might include grouping of design elements, linking precedence requirements (both technical and procedural), adapting to new technology developments, following required communication links, and accomplishing reporting requirements. The actions to be taken at each state depend on the prevailing product conditions. The nature of subsequent alternate states depends on what actions are implemented. Sometimes there are multiple paths that can lead to the desired end result. At other times, there exists only one unique path to the desired objective. In conventional practice, the characteristics of the future states can only be recognized after the fact, thus, making it impossible to develop adaptive plans. In the implementation of the DEJI model, adaptive plans can be achieved because the events occurring within and outside the product state boundaries can be taken into account. If we describe a product by P state variables s_i, then the composite state of the product at any given time can be represented by a vector **S** containing P elements. That is,

$$\mathbf{S} = \{s_1, s_2, \ldots, s_P\}$$

The components of the state vector could represent either quantitative or qualitative variables (e.g., cost, energy, color, time). We can visualize every state vector as a point in the state space of the product. The representation is unique since every state vector corresponds to one and only one point in the state-space. Suppose we have a set of actions (transformation agents) that we can apply to the product information so as to change it from one state to another within the project state-space. The transformation will change a state vector into another state vector. A transformation may be a change in raw material or a change in design approach. The number of transformations available for a product characteristic may be finite or unlimited. We can construct trajectories that describe the potential states of a product evolution as we apply successive transformations with respect to technology forecasts. Each transformation may be repeated as many times as needed. Given an initial state $\mathbf{S}_0$, the sequence of state vectors is represented by the following:

$$\mathbf{S}_n = T_n(\mathbf{S}_{n-1})$$

The state-by-state transformations are then represented as $\mathbf{S}_1 = T_1(\mathbf{S}_0)$; $\mathbf{S}_2 = T_2(\mathbf{S}_1)$; $\mathbf{S}_3 = T_3(\mathbf{S}_2)$; …; $\mathbf{S}_n = T_n(\mathbf{S}_{n-1})$. The final state, $\mathbf{S}_n$, depends on the initial state **S** and the effects of the actions applied.

Evaluation of technology

A product can be evaluated on the basis of cost, quality, schedule, and meeting requirements. There are many quantitative metrics that can be used in evaluating a product at this stage. Learning curve productivity is one relevant technique that can be used because it offers an evaluation basis of a product with respect to the concept of growth and decay. The half-life extension (Badiru 2012) of the basic learning is directly applicable because the half-life of the technologies going into a product can be considered. In today's technology-based operations, retention of learning may be threatened by fast-paced shifts in operating requirements. Thus, it is of interest to evaluate the half-life properties of new technologies as the impact the overall product quality. Information about the half-life can tell us something about the sustainability of learning-induced technology performance. This is particularly useful for designing products whose life cycles stretch into the future in a high-tech environment.

Justification of technology

We need to justify a program on the basis of quantitative value assessment. The systems value model is a good quantitative technique that can be used here for project justification on the basis of value. The model provides a heuristic decision aid for comparing project alternatives. It is presented here again for the present context. Value is represented as a deterministic vector function that indicates the value of tangible and intangible attributes that characterize the project. It is represented as $V = f\left(A_1, A_2, \ldots, A_p\right)$, where V is the assessed value and the A values are quantitative measures or attributes. Examples of product attributes are quality, throughput, manufacturability, capability, modularity, reliability, interchangeability, efficiency, and cost performance. Attributes are considered to be a combined function of factors. Examples of product factors are market share, flexibility, user acceptance, capacity utilization, safety, and design functionality. Factors are themselves considered to be composed of indicators. Examples of indicators are debt ratio, acquisition volume, product responsiveness, substitutability, lead time, learning curve, and scrap volume. By combining the above definitions, a composite measure of the operational value of a product can be quantitatively assessed. In addition to the quantifiable factors, attributes, and indicators that impinge upon overall project value, the human-based subtle factors should also be included in assessing overall project value.

Integration of technology

Without being integrated, a system will be in isolation and it may be worthless. We must integrate all the elements of a system on the basis of alignment of functional goals. The overlap of systems for integration purposes can conceptually be viewed as projection integrals by considering areas bounded by the common elements of subsystems. Quantitative metrics can be applied at this stage for effective assessment of the technology state. Trade-off analysis is essential in technology integration. Pertinent questions include the following:

- What level of trade-offs on the level of technology is tolerable?
- What is the incremental cost of more technology?
- What is the marginal value of more technology?
- What is the adverse impact of a decrease in technology utilization?
- What is the integration of technology over time? In this respect, an integral of the form below may be suitable for further research:

$$I = \int_{t_1}^{t_2} f(q)dq$$

where:

I is the integrated value of quality
$f(q)$ is the functional definition of quality
t_1 is the initial time
t_2 is the final time within the planning horizon

Presented below are guidelines and important questions relevant for technology integration.

- What are the unique characteristics of each component in the integrated system?
- How do the characteristics complement one another?
- What physical interfaces exist among the components?
- What data/information interfaces exist among the components?
- What ideological differences exist among the components?
- What are the data flow requirements for the components?
- What internal and external factors are expected to influence the integrated system?
- What are the relative priorities assigned to each component of the integrated system?
- What are the strengths and weaknesses of the integrated system?
- What resources are needed to keep the integrated system operating satisfactorily?
- Which organizational unit has primary responsibility for the integrated system?

The recommended approach of the DEJI model will facilitate a better alignment of product technology with future development and needs. The stages of the model require research for each new product with respect to DEJI. Existing analytical tools and techniques can be used at each stage of the model.

Conclusion

Technology transfer is a great avenue to advancing industrialization. This chapter has presented a variety of principles, tools, techniques, and strategies useful for managing technology transfer. Of particular emphasis in this chapter is the management aspects of technology transfer. The technical characteristics of the technology of interest are often well understood. What is often lacking is an appreciation of the technology management requirements for achieving a successful technology transfer. This chapter presents the management aspects of manufacturing technology transfer.

References

Badiru, A. B., Application of the DEJI model for aerospace product integration, *Journal of Aviation and Aerospace Perspectives*, Vol. 2, No. 2, pp. 20–34, Fall 2012.

Badiru, A. B., M. U. Thomas, Quantification of the PICK chart for process improvement decisions, *Journal of Enterprise Transformation*, Vol. 3, No. 1, pp. 1–15, 2013.

chapter nine

Technology performance economics

Jesse Brogan

It is no secret that living in the United States has become more expensive as time has passed. The shift from one-earner to two-earner households over the past 50 years can only be called a dramatic witness that the economy has become so degraded that it is threatening the welfare of the human family.

What has largely gone unseen is that this indicates a long-term and severe state of performance failure. The amazing truth is that modern economists are little concerned by this because of their reliance upon activity-based economic indicators.

The modern science of economics

The *why* is inherent in our modern science of economics, with its current focus on how an economy operates. The modern study of economics results from a close and effective observation of economic activities, with the collection and distillation of knowledge of its operation's causes and effects. This yields a science that is intent on predicting the impact that various economic stimuli have on an economy, supporting government's management efforts.

The weakness is made obvious by addressing the source of its knowledge. A study of past operations can only support doing what we have done in the past. Any change in the way that an economy functions is wholly outside the vision of this science. Modern economics is designed to balance and rebalance economic operations. It supports efforts that involve internal economic adjustments.

This limitation, arising from the very source of the existing science, takes on importance when we see the loss of benefit for people, where both parents must now work outside the home just to provide for the family's daily needs. We have significant failures in delivery of value to people.

Our deeper challenge is that we find ourselves doing variations on systems that are increasingly failing to serve us as people. The modern study of economics does not support economic improvement in the sense of reversing the loss to people; it just rebalances internal operations.

A new science of economics

The basis for a new area of economics is available through an application of engineering, starting with a purpose of systemic performance. That development will further require an application of scientific investigation in a new and very different direction.

The difference is endemic. The systemic approach starts with purpose instead of going directly to observation. It starts with a definable and measurable something-to-be-accomplished through the operation of the economy. Having a purpose supports managing the economy, with management defined as gaining something of value through the efforts of those who are managed.

Who accomplishes what?

The first step is a definition of the performance purpose, of what to accomplish. This addresses what the management effort should gain through those who operate our economy.

Our general answer is also found through performance engineering; it is looking at the inputs and outputs of an economy, while approaching the economy as an operating entity. The *who* will address those who provide the economy's inputs and receive its outputs.

The economy does what people decide to do. The foundation for our new direction is that people commit their life energies into an economy. As a complementary understanding, people are the ones who face challenges in what the economy is doing. Where governance of the economy is achieved through representation, people are also the ones to be served by any intentional economic management effort.

Our generic answer is that any performance purpose will be converting the organization's inputs into its outputs. The new science will approach the economy as an operating entity that performs a conversion and delivery process.

Treating the economy as an operating entity opens the way (Figure 9.1) to identify purpose-based owners, customers, and active operating elements. Directly or indirectly, people own their economy and everything in it. The owner of the economy is people as a corporate body.

People are the only party in interest

People are also the active element in the economy. It only does what people do. The resources consumed in economic operation are the life energies that

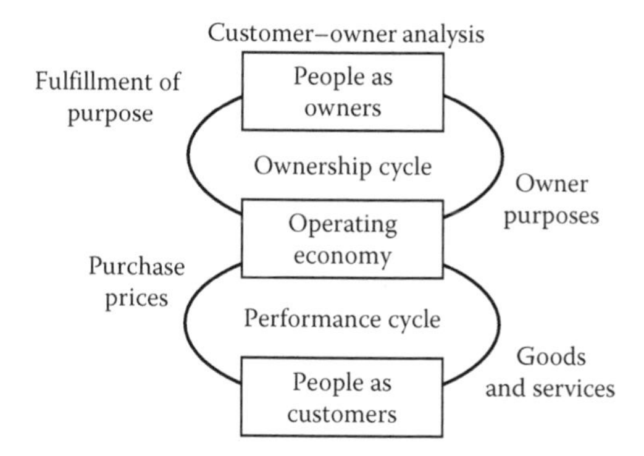

Figure 9.1 Customer–owner analysis.

people put into the economy and that is further seen to be what people decide to do as their part of being productive.

Customers are those with purchase decisions and are the ones who give an organization what it needs to operate, because they receive what they value from its operation. People put in their life energies in order to receive the goods and services that the economy produces and delivers to them. The customers are people who are potentially productive, those who commit their efforts to taking an active role in the economy in order to receive what it produces for them. Economy customers are also people.

The performance purpose for an economy is converting the life energies of productive people into what these same people receive and value.

Our new economics must be a people-oriented study addressing the value consumed from people in productive efforts and the value they receive when they contribute to their economy's operation. The result will be a science with the very purpose of supporting those who perform effective management of the economy.

Universality of people

One of our most pressing needs is to redirect the way people address each other. We-the-people are the corporate owner of all that is owned. People create and own all corporate bodies. There are no other ultimate owners. There are no people who are different than those who own the economy.

And here is one of the keys. The people who are dependents, like children learning in grade school, are the same people who will be young workers tomorrow. The people who are workers today will become retirees and have to live off their savings and investments. The people who are yanked out of the workforce by circumstance can be anyone and are not some defined class. The young become the adults, and the adult becomes the aged, and they are the same people at different ages.

There is no good way to separate these into classes and see to the interests of one of them at the expense of the others without pitting people against themselves. There is no good way to favor one age group without taking favor from them at some other age. There is no good way to favor owners over workers or workers over customers without causing damage to everyone.

People are the public

Even more on point, people are the public and there is no other public that is different than the people.

Family is also a customer of the economy

Family is just a crossover; it is where people take care of one another without regard to any other group affiliations. Being human, family is also universal, and family decisions on committing family resources gain family goods and services for the family.

There is no independent economic group such as *workers* or *management* who are to be benefitted at the expense of other people. For the new economics, there is only one party in interest, and it is people. There is no separate group that should get rich on the backs of others or that should be poor because others take away the fruits of their labors.

As an economic operating tool, wherever we find ourselves trying to see to someone's interest at the expense of others, we will be at cross purposes with the operation of the economy for the benefit of all people. The purpose for managing an economy is not served through corrective adjustments, but through a redesign to benefit all people.

Attempted economic adjustment to benefit those most in need takes up life energies and sets people at odds with one another, but produces nothing for anyone to value and receive. Such interference is always focused on fairness, not on prosperity, not on the welfare of all people, and certainly not on the effective and efficient operation of an economy as an operating entity that serves the people.

Universality of prosperity

Consider kids playing marbles, and the question is one of who has the most marbles at the end of the game. Some will win marbles in play. Others will bring many marbles with them and start out with more. Still others are not interested in playing at all and retain what they have. The question of a fair distribution of marbles may seem important for the moment, at least for those who choose to play.

Let us give each child a few marbles, so that they can enter the game. Then, each of them will have the opportunity. Since we don't have any marbles to give them except our own, perhaps we need to limit the marbles that others have. We will seize on them for the good of everyone, for then everyone gets to play.

Life is not a game of marbles

Prosperity is not in being able to play, and not everyone wants to spend their afternoon shooting marbles. Many are quite content with what they have; still others feel deprived because they would like to play and cannot. A few of the children will likely end up with most of the marbles.

People have life energies, wants, and needs. People are not different in this. In general, people have choices for the life energies that they would expend, or their parents have choices that impact on what they do as dependents. If one child decides to play marbles and another to study engineering, that is certainly their choice based on what each values.

Prosperity is not in the measure of how many marbles you collect, unless that is what is important to you. The individual provides the real definition of value, the definition of prosperity, and it is based on the personal viewpoint and understanding of the individual who makes that determination.

The common illusion of have and have-not prosperity comes from trying to measure prosperity without reference to the person's own sense of value and importance. Some children will be desperate to have all the marbles and count their prosperity in terms of this collection. Another child will consider going to the movies much more important, and getting there will be that child's definition of prosperity. In this, prosperity is both universal to people and individual to each.

Let the one who values the movie sell his marbles to the one who collects to get the last 10 cents he needs for the theater ticket, and both become prosperous. Both can receive what they value.

Engineering essentials

What we are developing is the hybridization of performance engineering with the study of economics. By applying engineering essentials, we will be opening the way for economics to serve those who have things to accomplish through managing the economy. The ones who have this purpose are the people who make up the economy and their representatives. This will be a new science that supports effective governance over the economy as a means to serve the people.

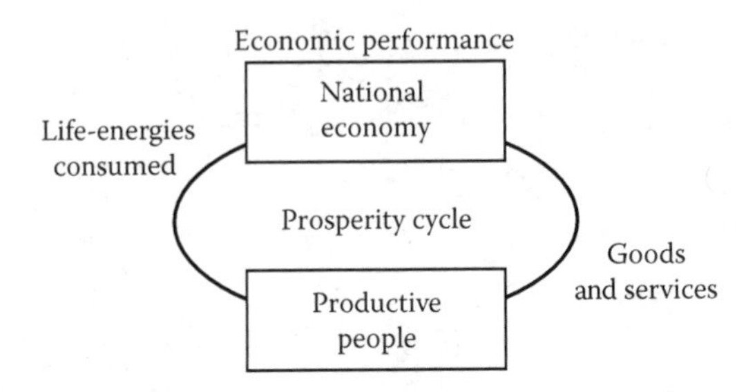

Figure 9.2 Economic performance.

Engineering provides technical support for those who have things to accomplish

With economics as a people-based science, efficiency engineering will support managers and leaders in operating an economy. The value of this new science (Figure 9.2) is remarkably greater than what we have because it maintains focus on a clear sense of purpose. When we have a purpose, with defined inputs and outputs, we also have a means to define performance. Economic value is then inherent in the value consumed and the value produced. When referenced to customer decision, it is the value that productive people and family members have for committing their life energies they commit to productive efforts, and the value that these same people have in the goods and services they receive because of their commitments.

Unlike the somewhat sterile focus of modern economics, one that deals with internal economic operations, our defining of input and output values requires the personal valuation of those who make the commitments of their life energies and who also receive and value economy products. Instead of leaders serving the economy as the basis for economic activity, the new economics will serve people as owners, operators, and customers of the economy.

The difference between the traditional and new economics is endemic and further effects the purpose for governing over the economy. Traditional economy management sees the economist as support for governmental investment in the economy, and governance includes expending resources upon the economy to have a positive result upon its internal operations. Our new economics serves a different purpose. It is minimizing what people have to put into the economy, while maximizing the value they receive from it. Expenditure of resources is addressed as personal cost and is in terms of a person's expenditure of life energies. Value is also addressed as personal value in what comes back to people as goods and services.

Further, this indirectly points out the degradation of our economy under guidance from modern economics. Benefits from rebalancing the way the economy operates rarely serves all people. With traditional economics, personal evaluation is not even a central consideration. Modern economic adjustment requires resources for change, but generates little if any return to people as customers of the economy. Application of the traditional science encourages a slow decline in services for people.

From the history of industrial engineering

What we gain from using the logic and approach of industrial engineering, performance technology, is seen in what this engineering has accomplished in the past. It was instituted at a time when workers were so hostile to managers that they felt encouraged to restrict their personal performance in order to assure the jobs of other workers—a purpose that ran counter to the purpose for hiring workers. It was a competitive us-and-them environment.

Frederick Taylor, credited as founder of industrial engineering, writes of being threatened with bodily harm when he acted to redefine the purpose of workers to benefiting the operation of the business that employed them. Since that time, he has also been roundly criticized by senior managers for treating workers as cogs in the wheels of industry, even though that is what managers had been doing prior to his influence. Again in his writings, we see how Taylor promoted treating each worker as an individual, and promoting their individual welfare, the opposite of what management has accused him of doing.

The effect achieved by Taylor's approaches and techniques was a most significant increase in productive output, coupled with a reduction in the time and effort put forth by workers to accomplish that increase. It was a win–win result of such potency that it has become the basis for the modern foreman as a performance manager.

For this chapter, we are addressing a potential change from an internally competitive economy managed by balancing conflicting purposes, to a performance-driven economy that succeeds through an increase in the value that gets delivered to people. The techniques and approaches that increased performance in the production environment provide a good place to start when our purpose is gaining a like effect in the larger economy.

The recorded resistance to performance approach, both by the workers and the business leaders, is also a valuable lesson. Their personal hostility to the changes became ineffective on seeing beneficial results. We can expect the same sort of resistance to applying performance engineering in the current economic environment. We can expect the same sort of performance increases to overcome that resistance.

Performance engineering and economics

The heart of performance management is measurement. There must be a measurable difference between a successful and a failed action before the success can be managed.

Performance is measured by inputs and outputs

Our performance measure is necessarily a personal measure by those who are productive, by those who receive benefit from the economy because they commit some of their life energies to its operation. Our metrics are the value of the life energies people decide to commit and the value of goods and services that these same people expect to receive in response to their commitment.

Measuring this value is possible as the very decision to commit witnesses to an in-fact valuation that determines the resources used up in economic operations. When we measure what people decide to provide to obtain the goods and services they expect to receive, we are measuring the performance of the economy.

Harkening back to the original observation, if it is taking the work of two family members to raise a next generation, where it only took one some 50 years ago, we have a measure that indicates an open and obvious failure in the basic operation of the economy.

The concept of *efficiency* is the appropriate analytic tool, with efficiency determined by comparing the value of outputs to the cost of gaining those outputs. The very definition of economy efficiency is aligned with the customer's decision process. We can address the value of economy operation in terms of its efficiency in turning life energies into those goods and services that get delivered to productive people.

There is no such thing as absolute efficiency. Efficiency is a relative measure. There are no direct measures for efficiency, only for the values of inputs and outputs, and the relationship between them. Efficiency is improved by a reduction in the value that productive people put into the economy. Efficiency is improved by an increase in the value productive people receive from economic operation. Instead of measuring economic efficiency directly, we must measure increase or decrease in efficiency.

Blivet math

Due to the common use of nonengineering applications, we need to address economic capacity in a new way. I call this blivet math, honoring the limits and capacities that are inherent in addressing a whole economy as an operating unit.

Restrict any basic input, and output suffers

An operating unit has various internal performance processes. Each process is likely to have a requirement for various inputs if it is to be a success at meeting its purpose.

Consider a production effort to produce bicycles. We have a supplier who cannot deliver all the steel wire we need to make spokes for bicycle wheels. Accordingly, production of the wheels drops, soon followed by a reduction in the number of bicycles that can be sold to support further production.

Performance management assures that needs for basic resources are met. Performance reacts badly to resource restrictions.

Consider again that the supplier for spoke wire delivers twice as much of this material. It will not support making more bicycles, as the finished product has other parts as well, and all must be provided more abundantly before production can be increased.

Increasing the supply of any one resource does not increase outputs that also use several other resources.

With this understanding, an economy is not all that flexible in the short term. It cannot adapt to meet new requirements as they arise; change is more likely to result from a steady pressure to change applied over time. Any new demand for outputs is likely to require new resources before the economy can react, and resources come from people by their personal decisions.

Change is always treated as a cost

What this means to us as to new economics is that the short-term operation of the economy must be treated as a stable process with a stable consumption and stable output.

Giving some of that output to beneficiaries will reduce the output going to productive customers. Artificially increasing the prices (man-hour expenditures) for any particular output will decrease the expenditures that are associated with other products or will limit the products that get delivered.

Simply putting a new demand upon the economy, even if valid, does not guarantee a short-term supply of output to meet that demand. Changes to the inputs and outputs of an economy are likely to be disruptive, and economic reaction will take time.

The appropriate logic supporting change will be investment rather than simple improvement. There will be costs to cover expenditures to overcome before we (as owners or customers) can realize any benefits from changes.

For the design of change, we must always address costs, and payback of those costs, before the change is justified. In the wisdom of industrial engineering, intelligent improvements to the economy will be initiated incrementally, and not be continuous or ongoing.

Efficiency and the economy

Industrial engineering is the specialty associated with operating efficiency, and the lessons gained in the production environment can be applied to good effect to enhance the productive operation of an economy.

Eliminating waste is the most effective way to increase efficiency

We have a general diagram (Figure 9.3) supporting efficiency work. Type 1 inefficiency is found in the misuse of incoming resources, consuming them in ways that do not contribute to generating value that gets delivered to customers. Type 2 waste is found in the delivering output to beneficiaries, to noncustomers, or in creating things that never get delivered to the benefit of anyone. Engineering terminology addresses these losses as waste. Type 1 waste is waste of economy inputs, life energies used for nonproductive purposes. Type 2 waste is waste of outputs, delivery of value to noncustomers and creation of goods and services that have no customer value.

Cutting the cost of operation is a management technique preferred by modern leadership, but yields much less effect than eliminating waste, and it often creates unforeseen damages. The heart of professional efficiency improvement centers on simply ceasing to do things that create waste.

We find massive type 1 waste in areas such as security trading. Trading investments produce no goods or services that go to productive people. It is a way for people to get a living without being productive. Using blivet math, the general price of harvesting value from investments through trading them is the loss of income for those who invest to own elements of the economy.

The result of promoting trading has been loss in ability to earn through making investments—return-on-investment is now harvested

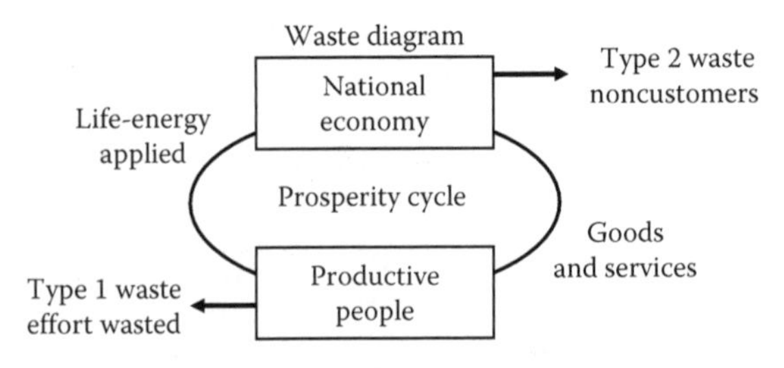

Figure 9.3 Waste diagram.

by traders. Our massive increase in investment manipulation has yielded a corresponding loss in ability to earn value from having investments. It all but forces investors into trading just to keep the value they have tied up in investments. It has become very difficult to save enough to retire comfortably—forcing people to work longer or face poverty.

Again, security trading increases what the productive person must commit to gain the benefits generated by the economy.

We find massive type 2 waste in entitlement programs that redirect value from those who are productive to those who are not.

Again using blivet math, we recognize that, without change, the economy only has so much goods and services that it is able to produce from the life energies it consumes. Any part of that performance result that gets redirected to noncustomers is not being directed to the customers, the decision makers who add to economy performance. It is a cost to be paid by productive people, a cost that does not contribute to what productive people get for their efforts. It is a loss of prosperity for working people.

One immediate value indication (an indicator rather than a true measure) is seen in comparing the value of the time and effort of people to the value of dirt-based products, such as farm land, road construction, oil, or precious metals. The supply of dirt is relatively fixed. The more life energies a man must commit to get dirt products, the less valuable that man is to the economy. This provides a snapshot measure for evaluating the success or failure of past economic operations. When a man must work a month to pay for 20 feet of roadway construction, the serious nature of our economic failure becomes painfully obvious.

Efficiency is a management tool, not a scientific one. It is applied here because it is aligned with the purpose for having a new economics. Engineering serves the personal prosperity purpose, and it is that purpose which will drive the development of the new science.

An application of economic engineering will eventually have value and be set to optimize the operation of the economy for the benefit of people.

Defining The People

The people is not the same as *a majority of people*. *The people* speaks of an effective unity among people.

The very concept of an effective unity has not been addressed in modern economics, but performance engineering does have a means to approach such a unity. The source is the Pareto principle, a statistical idea from noting that economic value addresses people—and people are statistically normal.

When in agreement, 80% of the people can speak for the people

The Pareto principle (or 80–20 rule) is that a 20% disagreement with anything is statistically normal. If there is greater agreement than 80%, it is effective agreement among people and they can act as a unit. If there is less than the 80% agreement, it is a popular understanding rather than an effective unity.

For social or political applications, 80+% agreement will be able to act without effective opposition; the remaining people in disagreement will continue to resist as individuals, rather than having a regular capacity to coalesce as a group in opposition.

The most immediate application will be people determining value through their decision to commit life energies to the economy. If something is not considered valuable by 80+% of the people, it is not something that people value as a unit.

For the purpose of management, only those things that serve 80% of the people will support directive management action on behalf of the people.

Group performance

Performance is almost always accomplished by people working in groups as opposed to singular efforts. This is as true for an economy as it is for any other productive effort.

From athletics, when a high school team competes with a sand-lot team, the structured and practiced team with a coach wins. The difference is an intentional and centralized point of management and the resulting ability of the players to rely upon each other to support what they are doing. When the sand-lot team meets a group of neighborhood children that are just out for themselves, each trying to do it all, it is the sand-lot team that wins.

And then we can address the equivalent to a balanced economy, where each part is urged to see itself in competition with other parts: management against labor, older workers against youth, and regulators against performers. There is no less effective approach. When the success of one is found through denying success to others, then all that can be accomplished is what the others cannot prevent.

There is no less effective structure than people gathered to oppose each other

We have defined performance in terms of a conversion process. People working together for a universally valued performance result will be many times more effective than people who are independently trying to do what they value, competing with each other over what they are to accomplish.

Success, as we have defined it for the operation of an economy, is corporate success for the people, a success as valued by 80+% of people. In the sense of economy purpose, success is wherever 80+% of people accept that they have a reduced expenditure of their life energies on the operation of the economy, or 80+% of people see that they receive more of what they value for what they do commit.

Arranging for some people to gain their success at the expense of other people is a blind alley; it goes nowhere. Shifting work or wealth around accomplishes nothing in terms of the overall welfare of the people, or their corporate prosperity. What is worse, this very concept of rebalancing the economy is based on serving and maintaining internal competition where the success of some is gained at the cost of others. It assaults our performance purpose.

One challenge is management of the economy, gathering resources for a productive purpose. This includes bringing people to see the universal benefits of increasing what an economy can do for people, and reducing what people must expend of their life energies to gain what they value.

High-level performance is an answer to that challenge, and it comes when people see shared purpose in what they do. People work together when they have common purposes, such as reducing what it costs to operate the economy or increasing what people receive for the personal resources they commit to productive efforts.

Alternative value sources

With modern economics, we have many impersonal sources of value, as with basic resources. These are considered drivers for value in rebalancing the internal systems of the economy.

This balanced approach to value can be seen as a major cause of the loss of value in the time and effort of people. Attempting to preserve resources as a cause is unconnected to personal valuation; it relies upon artificial value assignments.

Where dirt becomes more valuable relative to the person, the person is less valuable relative to the dirt

With our new economics, people are recognized as the only true source of value. Basic physical resources have value only as people value them. What people do not value as individuals can have no real value to the public.

When it comes to the value of people, only productive people add value to the economy, and they do so when they commit their life energy in ways that contribute to economic output. Other people, including the young, old, ill, or infirm, are consumers who add nothing back. They are not customers but beneficiaries.

When a person acts as part of a family, the family purpose includes providing for its unproductive members. Caring for unproductive people is not appropriate as an economic driver. When a person is out of work, or engaged in efforts that do not contribute to economic performance, they also become beneficiaries rather than customers. They do not have purchase decisions that will commit their life energies to performance activities.

As is seen for children during school years, they are potential resources that are not being realized; they are a cost from which no benefit is immediately generated.

Beneficiaries do not contribute to the operation of the economy and do not have opportunity to generate value for others to consume. As to the general performance diagram, they represent type 2 waste.

In the wisdom of efficiency engineering, it is intelligent to minimize the cost of nonproductive people. The more the economy expends upon them, the more waste it generates for the economy as a whole.

It is instructive to address this as to children of the family. In economic terms, children are intentionally nonproductive, and cost should be minimized. In family terms, these children are the very purpose for family. It is the family that should be caring for children, not the economy. The proper source of economic valuation of children is for those in the family who commit their life energies to get goods and services for the family.

This is a primary understanding. The family is the effective customer, and the unproductive children are internal units in the family. It is then the family that provides valuable life energies to the operation of the economy. Those working members of the family are the ones who do the valuation of the time and effort they (as part of the family) decide to commit, and value the economic goods and services that the family receives and consumes. This is a purchase decision; it defines an effective working relationship between the economy and the human-family as its customer.

Caring for nonfamily beneficiaries

We also find ourselves addressing the challenge of those widows, orphans, and retirees who have no family support, a general statement for those in need of public support because they are not directly supported by productive people.

The question tumbles back to defining *who* should be responsible. The economy, with the purpose of serving productive people, is obviously not the right entity to care for these. It is people who value other people. People will make personal-charity decisions; they will value the care that comes to those who are unsupported by family units.

Any public attempt to redirect output to these unfortunate human beings will take it from those who are productive for the benefit of those who are not. Public charity is always in defiance of the very purpose for managing an economy.

On the other hand, private charity is a valid expenditure of what people earn and is then prioritized right along with meeting other personal needs and wants.

The problem is seen to be a human limitation. The answer must be from a human capacity. Even as people can choose to act as a unit, the economy can be used to value support for these unfortunates as a public purpose—but that requires 80+% agreement on the level of support that is provided by *the economy* when it acts as a unit.

The other answer is personal, and this is the driver for those who see a higher level of support as having a personal value. It is up to them, using their resources of life-energy expenditure, to provide personal contribution to the output that goes to the needy.

This retains both public decision by the people as to a base level of charitable support and personal decision by productive people as to any greater charitable-value delivery. In accord with the purpose for the economy, this valuation of beneficiaries will then be based on the value decisions of people.

We also need to address economic interference by those who govern for the *good of the nation.* This returns to the basic question of whether it is possible to have a public purpose that is not a purpose of the people. The answer is that representatives are only empowered to see to the will of the people they represent and not to seek out and serve some special (but different) public. Those efforts that are not valued by the people as individuals cannot have public value.

Government management over the economy

As a general statement, people are the only party in interest. There are no other decision makers, no other owners or customers whose decisions and actions determine the operation of the economy.

Corporate entities are also owned by, operated by, and are producers for, people. Even government is there for people. In the United States, this representation is the effective reason for having a government.

As a hard lesson, if the people have to be compelled to action or inaction, it is not a valid economic decision on behalf of the people who are compelled. People cannot be removed from the determination of value without removing even the possibility of public purpose, with resulting loss of operational efficiency. The term for a government's non-representative impact on the economy is *mismanagement*.

As far as the economy is concerned, the purpose for government is management of the economy for the benefit of people; it is converting the life energy that citizens commit into what the same people value.

Representatives are not parties in interest except as they represent the people. If a representative would represent some people over others, then he or she is not representing the people as a corporate body, and not representing all the people as individuals.

Further, representative leaders get to represent the people, but not to do evaluation for them, or on their behalf. The people are the only ones who have purchase decisions based on the value they attribute to economic inputs and outputs, value to them as individuals or families. They are not obliged to honor the economic decisions of other people through their private commitments, and those private decisions determine the operation of the economy.

Directing people to make specific decisions is unauthorized by the people

Enforcing government management decisions that the people do not honor is a source of competition between government and the people it is to represent. It is like a football coach who would deliberately throw a game. Our current economic competition between personal decision and government direction causes waste.

This emphasizes what our new economics is to accomplish, the effective management of the economy for the purpose of delivering benefit to people. That only happens when government also serves the people through its economic management efforts.

Alternative investments and the economy

Effective management through government will always involve a personal investment by customers. This has obvious potential when addressing noneconomic value areas such as health, welfare, and safety.

These have human value. It is not an inherent value that justifies a mandate to action, but it only has value as people recognize that value. People are the necessary decision makers who will determine what value it has to them, relative to other things that people value, such as food and shelter for the family, education of their children, care for elderly family members, entertainment, and the like.

Modern economists have resource-based economic concerns. The new economics addresses the performance of the economy, and nonhuman resources are addressed as being without inherent value. They are given value only as people value them. Preserving resources is not a valid economic purpose. Any valid government preservation of resource must include serving people through the preservation, and it must be the people who recognize that value. The economic value consumed still has to come from people. The value produced still has to go to productive people.

Any resource-based proposal is then subject to economic investment decision, where the cost and value to people are compared, and people either support or refuse to support the proposal.

Regulation of people is subject to investment analysis

If significant numbers of people need to be directed to act or to avoid action, that regulation will not be the result of their valuations. Regulation of people almost always accomplishes economic mismanagement.

Regulation of corporate entities is easier to justify, as they are not evaluators of what the economy produces. Corporations must go to their own corporate owners and investors for valuation.

The personal prosperity index

To establish management over the economy, we need performance metrics to measure the success or failure in its operation. This is the technical support that performance engineering initially defines for our economists, so that they have the tools to support government in its management of the economy.

Personal prosperity index-based governance

There is no absolute measure for prosperity; the personal prosperity index (PPI) is a relative metric. Accordingly, we start with an arbitrary number, say 1000 PPI units, for average income from a work effort committed during some selected period of time.

Income from investments is another consideration. To establish this 1000 PPI points, we will add equivalent average investment income, with the value of investments in terms of equivalent earnings-from-working-hours. The 1000 PP represents the whole income for the average person. It will have a component representing personal performance earnings and another for investment earnings.

We must also consider nonworking hours that are expended on a voluntary basis—which we know have value to those who expend them. People value their time in various ways, even as they value the time they work because of what they earn.

Our voluntary uses for our nonworking hours have hour-values equivalent to our working hours. If there is an imbalance where one becomes more valuable than another, we can assume that the individual will shift hours to restore that balance.

These nonworking hours represent PP units to add to those used to gain a salary or wage. This might, for example, add an additional 1500 PPI for 60 hours per week of personal time (sleep and other non-voluntary hours excepted), to give 2500 equivalent PP that can be expended by the average person.

Each major element of expenditure is to be categorized and associated with a contributory number of PP units. For example, there may be 500 PP associated with the time and dollars required to have and maintain a house or apartment (20% of 2500 PPI units expended). A working metric for measuring and relating prosperity to housing might be square footage of personal living space. A family of three in a 1200 square feet apartment would have an average of 400 square feet of personal space. The square-foot metric of 400 would then equate to a PPI contribution of 500 PP for each person; or 1.25 PP per square foot.

A person's contribution to economy performance is measured by the PPI relative to hours expended on productive efforts. This is the whole value of performance relative to what it costs to gain that performance. The sum of all contributions for elements of expenditure will be the whole number of points (2500 for this example). This provides the baseline for measuring relative economic prosperity, both as to the whole economy performance and as to the performance of the various productive and nonproductive elements that contribute to the larger measure.

To measure the PPI for another period, we apply the same nondollar metrics, such as the square footage in our example, in the new time period. On measuring this metric to be 410 square feet per person during the new period, we have a 12.5 PP increase in the housing contribution to the PPI, representing an increase in the value customers are receiving from the economy as to housing. We do the same for all categories, and we sum all the categories.

We also must adjust for any shift between working and nonworking hours, and recognize any change in average investment amounts, to derive the new time period's PPI. If the PPI goes up, we have an increase in personal prosperity for the average person. If it goes down, we have lost personal prosperity. Also, if the productive man-hours needed to generate a level of performance goes down, then there is a corresponding increase in personal prosperity due to gaining the performance result at less cost.

If personal earnings exceed expenditure, putting more into investments, this also represents an increase in personal prosperity; it measures the opportunity for people to increase future income through their savings and investments.

The PPI is the appropriate tool for governance, and developing and maintaining such a metric should be one of the purposes of governance as it will bring both leaders and the people to focus on how government serves its citizens' economic purposes. Tracking the PPI and its various components supports governmental management, and with publication, it also supports representing the people.

For each subcategory within the PPI, there is a corresponding PPI contribution, a number of PP units that represent the value people feel to be associated with that category. That value can grow or shrink—and increasing the value of the PPI contribution is a purpose of governance over each area of the economy.

The PPI can be used to track individual decisions to desired effects

The importance of this is that every potential action affecting the PPI contribution becomes separately manageable. There is a difference in the success or failure of the economy as to each tracked area based upon the political activity that is contemplated or accomplished. Changes to the PPI contribution mark a success or failure for every act of economic governance.

Even more on point, these are elements of personal evaluation. The same people who elect leaders are the ones who will feel any increase or decrease in expenses or in what they receive as a result of changes in their expenditures.

The maintenance and use of the PPI is an act of economic management and should be the responsibility of those who govern. This should include publication of expectations as to PPI impacts, as these will be indicative of the impact associated with any actions of economic governance. That publication provides a professional economic support for the people, measuring the quality of governance being provided.

There is also the opportunity to evaluate the acts of governance that are noneconomic, those proposed or undertaken based on health or welfare. This addresses cost and benefit in a way that uniquely relates to human wants and needs that are not just economic, but are the essentials of personal prosperity. It provides investment information where the public can approve or reject proposed actions based on their valuation. There will be corresponding respect for leaders who increase value and loss of respect for leaders who might impose unwanted investments upon the public.

PPI challenges

There are immediate questions that will need the attention of the economic scientist. In establishing the PPI, do we address the life-energy expenditure on productive efforts as a negative (customer orientation) or positive (economy orientation)—what viewpoint do we take? Are we to address this life energy as a human expenditure, or as the incoming resource for economic activity. This, of course, will impact on how we address nonproductive efforts such as entertainment. Such human activities have value, and satisfaction can draw time and attention from other potentially productive efforts, even as it refreshes the individual supporting further work efforts.

Again, there is the question of the Pareto principle. Are we dealing with PPI measuring average individuals, or are we addressing the corporate people, such that 80+% agreement levels are needed to establish changes as beneficial?

This will impact on whether a significant benefit for a relatively small portion of people can constitute a meaningful economic improvement.

Is it possible to have true quality-based determinations when we are addressing individual self-referent metrics? These would seem to measure only what people have relative to what they can experience directly or through empathy/sympathy of others.

Can we address family as a true unit of humanity, or must we deal with people as individuals, and assemble our understanding of family from this?

And then there is the variability of humanity, the ability to learn and gain new experiences that have effect on the very basis for human evaluation of results. It is, for example, quite possible to have an outcome almost everyone initially values, but that is found to be prohibitively expensive in life energies in other areas when people act to bring it about. Their shift in viewpoint will not have effect on the baseline measure, but can significantly impact on personal valuations of the current time period.

Is this acceptable, or should other provisions be made to back value the baseline, possibly contaminating its initial valuation to provide a more effective comparison.

Money

Traditional economics, being focused on activity rather than result, alternately treats money as a value of exchange and as an economic lubricant. The maintenance and handling of money has been approached as a major economic driver for achieving balance.

In the new economics, money is not a value in itself; the only source of value is people. Money rather functions as a negative marker for the flow of value to and from people. When people expend their valuable life energies on the economy, money flows back to them, and that same money provides a means to procure a corresponding value in goods and services. Should they turn that money back into the economy, it purchases goods and services that they can value. That money flows away from people as the value produced by the economy flows to them.

This viewpoint aligns our vision and understanding of money with the purpose of the economy, which is itself in terms of these two flows of value between the people and their economy.

The PPI measures the money earned by productive people expending their life energy and the use of it to procure what people value from the economy. Money is not a measure in itself, but indicates value as a percentage of earnings or expenditures.

Units of money

Traditional economics does not have a performance purpose for money; it only has need for a viable measure of exchange value. This serves the purpose of maintaining the economy's internal operation. Managing money also provides a means to stimulate or suppress selected internal economic operations; it serves rebalancing efforts.

To be optimum in effect under the new economics, money must maintain value, and the only source of value is people. Money will have to relate to the expenditure of life energies if it is to be in line with prosperity-based management. Life energies are in limited supply for any person and are expended on the economy in accord with the individual's value decisions.

The optimum money will measure man-hours of productive effort and will probably be self-referent to the whole expenditure by people upon the economy. The reference would not be a *man-hour* value, though an average man-hour value for the economy as a unit is a good basis for establishing monetary units. The value of the person's expenditure would then be relative to the contribution that the person makes through life-energy expenditures. It would relate to the average value of a man-hour committed to productive purposes. A person would be able to compare the value of their efforts as contribution to the like value of others. A person would be able to track their progress in contributing to economic performance as the value of their efforts increases with experience.

In yet another direction, the value of people to a nation would be inherent in the purchasing power of a man-hour of effort. There would be a measurable cross-national comparison of the value of people to governments. The comparison would be an open witness to the ability of the various governments when it came to economically supporting their people.

This average man-hour value is likely to be a very stable and would resist manipulation by either the people or the government. Inflation or deflation of the value would require increase or decrease in what people are able to contribute—and that would reflect changes in personal prosperity. A person's personal value to the economy is expected to increase as the PPI increases.

Governance of this money is much simpler than trying to use commodity-based money such as a gold equivalent. In the new economics, commodities have no real value and could not effectively represent value other than through valuation by people.

The new economics money would even provide new measure for the value of investments, how many equivalent man-hours we add through contribution to ownership of the elements of the economy, and what value is assigned to that investment in terms of working man-hours.

This new money would also be manageable in terms of new man-hours committed to productive efforts. The ability of corporate entities, even banks, to manipulate the value of this money would be minimal.

Governance and money

The basic metrics used to track the PPI are also useful for managing money. The available money supply would have to relate to the value of man-hours being added into the economy as it continues in its operation.

We can publish the impact of governance actions on the PPI; it could be regularly posted. The impact of various actions of governance would then not only have direct impact on the PPI and show the investments made in the name of the people, but would also show any undue or unusual impact on the money supply.

Money can be managed as a way to value people, goods, and services. The value people commit to productive effort is the basis for managing the economy. The value the economy delivers to people measures how well the economy is serving the nation and its people. A highly stable value of money in the economy can serve both.

Money and wealth

Wealth is not the same thing as an accumulation of money. Money has value only in transaction, in selling what has value or buying value. The value is in what is bought and sold, rather than in the money used.

Wealth is the accumulation of value. It is more aligned with prosperity than with money itself. Wealth just has a money equivalent, should it be transferred to someone else.

There are four elements of wealth. The first wealth is the source. It is having life energy that can be committed to productive efforts. Then there is wealth represented by money, by the marker that comes to a person in response to their delivery of their life energies to productive efforts. The third is wealth in the more durable goods and services that are gained through spending that money and addresses land, houses, cars, computers, clothing, and other things that people value. The final wealth is investment. This is value that a person commits to ownership through procuring investment resources, including stocks, bonds, savings, retirement, insurance, and the like. For valuing these, someone else will pay money to the holder if the provider gives up ownership. Having investments is entitlement to collect a debt that the person is owed because they invested some of their man-hour-based personal income.

For engineering purposes, there is no limit on wealth in terms of investment, though wealth in personal ownership is often temporary due to changes, as in a home growing old and losing some of its functionality, or needing additional commitment of value to maintain it.

For the philosophically inclined, there is also indirect wealth in opinion or reputation. There are ways to harvest other value sources more effectively when people trust one another or have value in other people. In that sense, even being part of a family has value, though it may or may not be addressed as wealth depending on the personal determinations of individuals and on their individual life experiences.

A few initial applications

Performance engineering is the art and science of practical problem solving addressing cost and performances. When it comes to improving organizational performance, it begins with the elimination of waste.

In an improvement effort, the engineer faces the intrusion of non-performance thinking. The answer is always guidance to refocus on a purpose provided by those who are most involved. The redirection for economy actions must always be to the personal decisions of those who count the most, people as customers.

The purchase decisions of people are determinative. They count more than any decisions by those who are only beneficiaries and more than decisions by special people who seem to operate the elements of either the economy or the government that manages it.

Harvesting from stock trading

Consider the type 1 waste of harvesting investment through manipulation of securities. The appropriate governance action is to block this way of earning income, using recognition that it takes from those who own the economy, removing their ability to earn from investments, and forces trading activities upon common investors just to retain the value of their investments.

Effective governance is available through heavily taxing unforced and short-term security trades. This takes profitability away from those who would trade securities as a source of wealth; the taxation is to essentially ban security trading as a way to make a living.

The deeper understanding, of course, is that this restores the proper and effective purpose for people who make investments; and that is to earn a return. This taxing approach eliminates efforts that are well paid, but generate nothing for people to value or receive. The former traders would have to find productive work to earn a living—and that would add to what the economy produces for people instead of simply harvesting benefits. Those who have invested to be owners will receive the appropriate increases in prosperity associated with return on their investments and not have to engage in further trading to maintain the value of their investments.

In a very real sense, it is the security trading system that is unfair, allowing some people to benefit from the productivity-supporting efforts of those who invest in the economy as owners. People will still be able to earn from investments, as operation of productive elements of the economy will still earn just as much. Those earnings will just flow to those who invest for a return, instead of those who manipulate investments.

Social security

As with most social programs, the current benefit is for some people at the expense of others. This program sets people against one another. The correction is not termination of the program, but creation of a program that effectively and efficiently serves the people. Value comes from investments, instead of creating economic waste by redirecting the flow of value to beneficiaries as entitlements.

There is no good way to make some people entitled to what other people earn

The goal is a systemic approach to caring for people in their later years, harvesting the benefit they gain when younger and more able to contribute to economic performance.

The appropriate direction is promoting savings, not establishing entitlements. The program will be based on supporting savings in a way that will support the operation of the economy.

The program can be used to share risk—but not through entitlements. We have commercial insurance for that purpose, and the person can buy the valuable economic service of shared risk at their own expense, their own purchase decision using their own sense of value. Government has no money of its own and should not be expending the resources earned by others unless it is representing those who earn.

There is no representation over the objection of the one who is supposedly represented. Ongoing representation is a matter of supporting the widely varying values and decision of those who are represented and is not imposing system requirements upon them.

In the replacement program, there would be accounts for retirement that are kept by regular commercial holders, and possibly guaranteed by government action and funded for that purpose. These will receive part of what people earn when they are productive and would be available to help support them when they retire. Unless they spend upon insurance or the like, they receive what they have put in plus any interest earned through the fund's ownership of the elements of the economy.

The difference from an entitlement system is endemic. Those who manage the system earn by providing a service valued by those who receive it. Those who choose not to participate will not receive the benefit. Those who never participate will not have their hands in other people's pockets, but will have to rely upon charity—that is the choice they will have made.

The government will no longer be in the business of stealing from some of its citizens to provide public charity to others. The decisions of people will guide governance, instead of leaders cutting the people out of the decision process in how they spend their own earnings.

Regulating economic behavior

Trying to regulate the behavior of all people is clearly beyond the powers of representative government.

People do not have to be ordered to do what they want to do

We can also have some effective public regulation. It is very likely that there will be regulation to prevent the intergenerational accumulation of wealth, as that would support a class of unproductive people. The proper regulation will not prevent the passing of wealth as inheritance, but assure that receivers are not encouraged to be unproductive. The key, of course, would be 80+% of citizen being in agreement upon any limits set on inheritance.

There would also be natural limitations, as where a person owned a family business, as with a farm or fishing boat. Such would likely pass a life estate only, but without inheritance taxes. It would then continue to be available for further passage to heirs without taxation so long as the one who received the inheritance continued it as a business. They would have to pay any inheritance taxes before they could transfer it to some other use or to a new owner.

The limitation of inheritance would likely be placed on accumulated wealth, such as entertainment systems, bank accounts, investments in other people's businesses, and the like. There will probably be support for inheritance taxes that will refuse great wealth to those who inherit.

In reaction, the planning of estates would likely shift to arranging productive positions for those who inherit, providing family business ownerships and preparing the next generation to assume effective charge over business resources. It would assure that those who inherit have the capacity to continue wealth through what they do.

Taxation

Things like general taxation of income would likely be replaced with public investment, with contributions that are supported by 80+% of the public because of what the governance would accomplish for them. That would be either in the value of government itself or specific and personally valued goods and services that people receive, in other words the U.S. Constitution states "…nor shall private property be taken for a public purpose without just compensation."

If backed by expectations of PPI impact, the specific acts of governance would be subject to public support or refusal. Those which had the 80+% support would be effectively mandated. Those which had less would be handled either by limiting costs to those who see value or by addressing the investment on noneconomic grounds, as would be possible in taking public responsibility for some level of damage recovery due to criminal acts. Giving that support might be a non-public benefit, but public support could be provided where the subject had 80+% agreement on a benefit that people value as exceeding the cost.

Summary

Our traditional economics study is no longer sufficient to support our needs. There is a real purpose to be served by a new economics.

Performance engineering provides an alternate technology approach, focusing on a different purpose than what economics now serves. Performance engineering does provide a means to see the general shell of a new economics study, and it is able to address some of its limits and directions of application.

The future need is for economist action, for the scientific community to reset their own expertise to deliver value to the people, value that is not supported under the current study.

Engineering is not competent to do the science; it is a problem-solving specialty that defines the economic challenge and provides a direction for solution.

The true value of result is realized through hybridizing the focus and perspective of engineering with the appropriate scientific application. It will center on the new development and organization of economic knowledge that fulfills the human purpose.

This chapter lays out the purpose and general directions to be pursued through a new people-based economics. Engineering may still be of support in the development of projects that implement this science, but the focus is ready for a shift to economists as drivers. Their efforts will do the science and that will gather and organize knowledge of personal economic performances for the purpose of predicting impacts on the personal prosperity of people.

chapter ten

Manufacturing competitiveness in Africa

Pius J. Egbelu

My people are destroyed from lack of knowledge

Hosea 4:6

Introduction

On a global scale, the manufacturing output in Sub-Saharan Africa is the lowest of any region in the world, except the Middle East. Similarly, Sub-Saharan Africa ranks at the bottom of all regions in contributions to global trade on manufactured products. Sub-Saharan Africa continues to rely mostly on the export of raw materials and commodities to support their economies. According to a KPMG report on manufacturing in Africa (KPMG 2014), instead of growing and expanding the manufacturing sector, since 1996 the number of manufacturing enterprises in Sub-Saharan Africa has been shrinking. Even among the manufacturing enterprises that exist, most are at the small scale level and lacked the capital, skilled labor, and technical knowledge to compete at the global level.

There is a direct relationship between manufacturing productivity and economic competitiveness. No country in the world has ever become industrialized without first promoting and engaging in active manufacturing. Furthermore, none has sustained and retained its industrialized status by neglecting the sector. Active engagement in manufacturing has been the right of passage to economic growth and development for emerging economies. The universality of this requirement suggests that African countries will not be the exception. Sub-Saharan African countries cannot expect to build and sustain robust and healthy economies without first strengthening their manufacturing base. Even with currently poor availability of infrastructure to support a vibrant manufacturing sector, with abundance of raw material, cheap labor, and closeness to major world markets, Africa is in an enviable position to transform its manufacturing base and become a player at the global level in manufacturing export. There is a direct link between manufacturing export and economic success. Manufacturing creates jobs and jobs create wealth. By engaging in manufacturing, African countries will be exporting value-added products with more stable markets and less price fluctuations than are typically experienced by the export of raw materials and commodities.

Causes of low manufacturing activities in Africa

Several factors contribute to the low manufacturing engagement in Sub-Saharan Africa (KPMG 2014). First and foremost is that the countries in the region, except South Africa, lack comprehensive policies that promote manufacturing to accord it the importance it deserves as an instrument for building the economies and improve their export earnings. The region has primarily remained an exporter of raw materials and has not embraced the idea of engaging in manufacturing as a value-added process to their exports. Weak legal and governance systems and high levels of corruption complicate the business climate in the region. These factors create difficult business conditions that inhibit foreign investment. Investors want to invest their funds in countries where they are confident the legal system is developed and matured enough to protect their interests.

High taxes and lack of access to technical and skilled labor force contribute to high production cost that makes it difficult for companies to make profit. Technical manpower is often imported and this raises the cost of doing business. The high cost of doing business weakens the region's global competitiveness. In an industry where there is fierce global competition, high cost of labor makes it difficult to manufacture products at costs comparable to those of other regions. Another factor that seriously affects African manufacturing competitiveness is the lack of adequate infrastructure. The region lacks the basic infrastructure that is needed to support a thriving manufacturing economy. Many countries still suffer from low access to energy such that even the few companies that operate in the region have to rely heavily on power generators as their primary source of stable power to run their factories. Power from generators is expensive and dirty; therefore, they cannot be used to operate sensitive equipment. The unreliable availability of power is strongly linked to the region's high poverty level as it inhibits the startup of new manufacturing enterprises and other business ventures that rely on electricity to function. There is no question that the lack of reliable power sources stifles innovation and creativity and consequently, entrepreneurship and the development of new manufacturing ventures. Furthermore, the lack of good road and railroad networks considered essential for commerce provides for inadequate infrastructure to handle the demands of today's rigorous and responsive supply chain systems. Efficient transportation infrastructure is critical for developing effective and responsive product delivery systems from suppliers and manufacturers to customers. Efficient transportation system is the backbone for global competitiveness in today's market economy. Customers are looking for suppliers who can deliver products with high value cheaply at their doors and this can be enabled only by efficient handling systems. The continued negligence and the underdevelopment of the basic infrastructure necessary to support the manufacturing sector make it all the more difficult for manufacturing activities to flourish.

Often not strongly emphasized when discussing the low manufacturing productivity in Africa is the role governments can play in promoting industry-based research and technology exchanges between universities and industry to drive innovation and productivity in manufacturing. Success in the manufacturing sector requires constant infusion of new knowledge to drive innovation and creativity in the processes used in product development, manufacturing, and delivery to market. Throughout the rest of the world today, companies are increasingly collaborating with universities to obtain innovative ideas to drive their businesses. Universities are seen and integrated as key players and sources of knowledge required to spawn new business and innovative production processes. While partnerships between university and industry are common practices in all industrialized and emerging economies as tools for economic competitiveness, such collaborations are lacking in African countries. Recognizing that universities are power houses of knowledge and that knowledge is inseparable from wealth creation and development, African countries need to emulate their counterparts in the industrialized and developing economies to more proactively and effectively drive their manufacturing sectors through technology exchanges that are enabled by enacting policies that promote university–industry partnerships. Such partnerships are beneficial to both sides and can contribute to the building of healthy manufacturing sector in Africa while simultaneously stimulating learning, research, and knowledge creation at universities.

The remainder of this chapter is focused on this very issue, that is, the promotion and use of university–industry partnerships as instruments to encourage and incentivize technology exchanges to stimulate innovation, creativity, and productivity in the manufacturing sector in Africa.

What Africa needs to increase manufacturing activities

As defined by the Council on Competitiveness, partnership is a cooperative arrangement or legal agreement between companies, universities, and government agencies and laboratories in various combinations to pool resources in pursuit of a shared R&D objective (Council on Competitiveness 1996). While we accept this definition because of its alignment with the spirit of the intent of this chapter, we also recognize that other definitions exist and that all partnerships may not involve research as will be seen later in this chapter. However, all do involve some knowledge sharing. Therefore, in this chapter, we also adopt the more broader definition that partnership is any arrangement between universities, industry, and/or government to pool their resources in promotion of technology exchanges through shared knowledge generated through R&D research, apprenticeships, personnel exchanges, training, and any other instruments that lead to the same shared goals. With this broad definition, having students undertake an industry project as part of a requirement for course or degree completion would qualify as an instrument for technology exchange that benefits both the company and the students or university.

With the exception of South Africa, partnerships between universities and industry are not common in Sub-Saharan Africa. Furthermore, university research is almost decoupled from industry needs and the day-to-day life of the people. The role of universities is seen merely as that of educating students, often without any exposures to real societal problems. In such academic settings, students complete their university education with minimal or no skills that make them industry ready at the time of graduation. Worse still, nearly all have no idea of how to use their university-acquired knowledge to start a business because their education had been devoid of practical training and experience that involve university–industry interactions. This kind of isolation and silo mentality between the academe and industry in Africa has to change if the region is to break away from its near lack of participation in the world manufacturing export. Without growth in noncommodity/raw material export, African economies cannot be expected to improve to provide the kinds of jobs required to grow the middle class. Greater university–industry interaction is needed to infuse new knowledge and skills in the manufacturing sector for the economy to grow.

To change the culture of isolation between universities and the manufacturing industry, African governments need to recognize and carve out a role for universities to more meaningfully engage and participate in building the economy that goes beyond the current state of simply producing graduates, most of whom would have little marketable skills. There is need to enact policies that promote and reward university–industry partnership to strengthen the manufacturing sector, enhance the quality and relevance of university education, and increase the participation of universities in building and contributing to areas of national critical needs. In doing so, they (i.e., African governments) will be treading in the paths that have been taken by all thriving economies of our time, including such countries and regions as the United States, Canada, European Union, Japan, China, South Korea, Singapore, Taiwan, Malaysia, Indonesia, and Thailand. If this is done and the sector is grown, it will serve as a catalyst to stimulate the growth of other sectors of the economy such as agriculture, mining, banking, transportation, information technology, and education to achieve specific outcomes such as employment growth, wealth creation, and increased quality of life, similar to outcomes observed in countries that have embraced the concept.

A natural question to ask at this point is, what should be the nature of such government policies that should be enacted to promote technology exchanges through increased university–industry partnerships? The starting point for the governments is the recognition of the role of research in building national economies. Research is an investment in the future. Therefore, funding for basic and applied research at universities needs to be recognized as necessary and critical annual recurrent expense item in the national budget aimed at unleashing the creative and innovative power that exists at universities through the pursuit of research in areas of national need such as manufacturing, information technology, transportation, health care, medicine, engineering, national security, military, and other areas of the economy that are dependent on science and technology.

Following this recognition, next the governments need to identify the product types or the aspects of manufacturing (i.e., priority areas) for which they have competitive advantage. These priority areas are not frozen in time but may change from time to time as maturity is attained in some areas or market conditions shift over time. While the areas of competitive advantage across multiple countries may be similar especially at the initial stages, in general one would expect them to differ. With priority areas established, governments are to appropriate funds annually to facilitate the desired university–industry research collaborations. The funding should be institutionalized to ensure program continuity from year to year. This way, a culture of research leading to innovation, entrepreneurship, collaboration, and competitiveness will be born and sustained.

The enactment of university–industry partnership policies to promote research, innovation, and creativity would require the government in each country to create an agency or multiple agencies to administer the budgeted funds. The primary responsibilities of the agencies are to solicit proposals from university faculty and researchers nationwide and award funds to promising proposals through their universities or joint university–industry groups to carry out the research associated with the selected proposals. The funded research must be in topical areas that are consistent with the goals and missions of the agencies. In countries where the funding level is low, one agency with multiple divisions within it can be designated to oversee the funds and make award decisions through a competitive process that involves proposal submissions and reviews. Each division within the designated agency may focus on one or more priority areas as identified in the national needs. In countries with larger funding appropriation, multiple agencies may be established, with each agency given primary responsibility to manage a subset of the priority areas.

Agencies are not to allocate funds to universities unilaterally based on a funding formula. Instead, the funds are to be awarded through competitive proposal process in which individual faculty, groups of faculty from one or multiple universities, or joint university–industry groups may collaborate to submit a research funding proposal to one of the designated funding agencies whose mission best aligns with their proposed project. Proposal submission may be on a rolling basis or at designated due dates throughout the year. Once sufficient proposals have been received or a designated proposal due date is reached, a panel of subject experts is invited to review the proposals and make award recommendations to the agency based on predefined award criteria such as proposal quality, alignment to mission, potential contributions to areas of national need, impact, qualification of the research team, and the assessed ability of the research team or investigators to successfully carry out the project at their home institutions. Armed with the panel recommendations and constraints on available funds, award decisions to institutions are made in the names of the proposing faculty or investigators. Awards may be made to cover a fixed period of time, normally 1–3 years, and are generally nonrenewable. Research goals and deliverables in each year are to be clearly mapped out in the proposal to help reviewers properly assess proposal outcomes as well as help the agency assess progress as the research is done. The competitive distribution of the funds ensures that the best set of proposals is funded during each review cycle, thus maximizing the potential cumulative return from the government investment.

While it is necessary for agencies to establish their policies on how proposals are reviewed to avoid corrupt practices, bias, and conflicts of interests, no effort is made in this chapter to address such policies in any detail. However, the omission of such detail should not be interpreted to mean the unimportance of such policies. Given the corrupt business climate in many African countries, it is absolutely important that policies be clearly spelled out on how awards are to be made to ensure funds are properly accounted for and channeled. Also not covered in this chapter is the post award management process at the awarded institutions to ensure that research funds are spent as outlined in the proposals. Award management at the agencies and the awarded institutions is an important aspect of the research process to ensure that government goals are met and that funds invested are well utilized. Those interested in details on proposal review process and post award management requirements can learn such details from one or more of the many premier research funding agencies around the world such as the National Science Foundation (NSF) and National Institutes of Health (NIH) both in the United States, the National Natural Science Foundation in China, the Natural Sciences and Engineering Research Council of Canada, and the European Research Council to name a few. Each of the countries and regions mentioned has multiple agencies that support research to account for different national priorities.

To ensure that an honest effort is made and that supported research has industrial relevance, proposals should be required to clearly address the relevance and contributions of their research and how it fits into the national priority agenda. Proposal teams should also be required to show evidence that some manufacturer or company is interested in the proposed research. Such interest can be demonstrated by having the companies commit as partners and willing to invest or provide additional resources to support the research effort. The partner company's level of commitment to the research can be measured by the amount and quality of resources it is willing to contribute to the research effort. Company contributions supplement the funding from the government. Partner company's contributions can be in cash, in-kind, or both. With the commitment of the partner company, the total funding for the research becomes the sum of the government funding and the contribution from the partner company. Multiple companies can become partners to a research project. Such industry contributions must be documented and verified during the life of the research award to avoid fraud and misrepresentations. This way, the government is underwriting part of the cost of a research project whose outcome may directly benefit the operations of the partner companies. When this model of a three-way (university, industry, and government) collaboration is sustained over time and across multiple collaborations and industries, a culture of university–industry partnership will naturally emerge. The emergence of such culture will produce greater and continuing interactions between universities and industry and thus facilitate technology exchanges between them. Increased technology exchanges will facilitate the constant infusion of new knowledge and discoveries from universities to industry and vice versa to drive manufacturing innovation, efficiency, and commercialization while at the same time enhance the educational experience of students. Increased manufacturing efficiency and effectiveness combined with innovation will translate to greater global competitiveness, thus repositioning the continent economically.

Technology exchanges

From the presentation in section "What Africa Needs to Increase Manufacturing Activities," one can infer what technology exchange implies. Technology exchange refers to the process in which skills, knowledge, techniques, and technologies developed in one economic sector is transferred to another and are commercially exploited to stimulate innovation and creativity that lead to new product developments, processes, improvement in methodologies, or the creation of alternative approaches to existing methods. It involves taking new discoveries of knowledge, skills, methods, and technologies that are generated from scientific research and transforming them into new commercial ventures that benefit the economy and the general public as quickly as possible as a dividend from research, especially university research. Technology exchange is anchored on the recognition that knowledge is wealth and that university research is a major source of new knowledge and discoveries. Technology exchange is a process of creating wealth through innovation, creativity, and commercialization derived from knowledge gained through research or other partnership modalities between industry and academic institutions. Technology exchange procedures make it possible for results generated from scientific and engineering research made available to a broader group of people or companies to develop and exploit to produce new products, processes, and services that can be put to practical use. Technology exchange takes place when new knowledge, technology, or method is discovered or created and there is a party willing to take the new discovery to practical application. In its simplest form, technology exchange involves at least two parties, the party that discovered or owns the new knowledge (i.e., the party that conducted the research or provided the new insight or generated the idea) and another party (i.e., industry) that acquires the new knowledge through some legal framework to apply or commercialize it. In a more general sense, technology exchange would involve a research team (usually a university or collection of universities), an industrial team consisting of one or more companies, and the public sector or government (state and/or national), the latter acting as an enabler by providing all or part of the funding needed to undertake the research.

Technology exchange is a highly valued process well recognized, promoted, and utilized in the industrialized economies to ensure constant stream of innovation is available to industry. The governments promote and support the process by supporting university research paid for by the public (i.e., government) to stimulate innovation and new knowledge discoveries that could lead to new product development and services. In these countries, support for research is strongly linked to global competitiveness and economic growth at home. In the United States, government funding for research has a long history and support for the concept formed the basis for the establishment of research funding agencies such as the NSF, which primarily supports academic research in all disciplines of science and engineering; the NIH, whose emphasis is primarily the support of clinical research; and the Office of Naval Research, whose domain is on research with military applications. Other funding agencies also exist.

In the 1980s when the United States was experiencing serious market competition from Japan and other Far East countries, Congress went even further to set up policies that strongly encourage university/industry collaboration to work on problems of national need. One of the areas specifically targeted was manufacturing. This was made as a manifest by the Bayh-Dole Act of 1980 on university–industry technology transfer (Mowery et al. 2001; Nelson 2001; U.S. Government Accounting Office [GAO] Report to Congressional Committees 1998). This legislation targeted the transfer of research results from universities to industry to benefit the public through job and wealth creation as well as growth in the national economy. The act led to a surge in patenting and licensing of university-generated intellectual properties (Debackere and Veugelers 2005). The linkage between university research and commercialization through university–industry partnership was considered so essential that it led to the establishment of entities such as the Division of Manufacturing in the Directorate for Engineering at the NSF. The drive also led to the establishment of government-supported engineering research centers at many universities across the United States with each center focused on some niche area for which the country needs to build up its capacity. All government-funded research centers were required to have sufficient industry partners to demonstrate their research relevance. The science and technology capabilities gained from the centers and the contributions they made to the rebuilding of American competitiveness were so significant and transformative that research center establishment has today become part of university character in the United States.

Furthermore, to emphasize the importance of research to economic growth, all federal government departments (ministries) in the United States were mandated by law to set aside part of their annual budget to support basic research through what is generally known as Small Business Innovative Research (SBIR). The SBIR programs only support research ideas that have their origin from small business enterprises or individuals unaffiliated with universities that engage in research with the sole objective of investigating problems that have their origin in industry and the potential for discoveries that could be commercialized. Again, the main driver is to stimulate innovation. The Bayh-Dole Act was motivated by the declining manufacturing sector, loss in manufacturing jobs, and loss of global competitiveness in the United States brought about by the fierce competition from Japan and other Far Eastern countries such as Taiwan and South Korea.

In Japan, policies that support university–industry linkage for the commercialization of university research results were enacted through the 1995 Basic Law on Science and Technology (Lee and Win 2004). Similar programs are found in the European Union countries, Canada, China, Southeast Asian countries, and some countries in South America. With the exception of South Africa that established the US NSF equivalent agency, the Foundation for Research Development, there are no Sub-Saharan African countries that have any programs that exploit the knowledge base at universities to spur research innovation and creativity and the commercialization of university research through university–industry partnership. By not having policies in place that promote university–industry partnership, African countries miss out on an essential and critical pathway for technology exchanges and consequently on their ability to meaningfully gain from the wealth of knowledge resident at university campuses throughout the continent.

Training next generation of technical manpower and providing access to research equipment

Based on experience in the industrialized economies of the world, the contributions of university research to national economies are indisputable. The absence of focused research on national needs leads to stagnation in knowledge and a decline in innovation and productivity. Most manufacturing enterprises in Sub-Saharan Africa are small in size, family owned, and lack the resources and the infrastructure to build in-house research capacity of their own. Many even lack the understanding of how research can transform their businesses. Thus, without government incentives, most of these enterprises will not grow beyond their existing sizes if left on their own. In fact, one can argue that without government support, many of these enterprises will go out of business in the future as they find the global competition becoming too steep to cope. If that happens, the level of manufacturing in Sub-Saharan Africa would be declining even further instead of rising.

Government support for university–industry partnership will provide research access to African manufacturers. Without such incentive, many will continue to be starved of the knowledge they need to expand their operations. The inhibition to growth manifests itself in multiple ways, including lack of access to high skilled workers, inability to meet international standards on product quality, inability to provide product variety, high rising production cost, and inability to design competitive supply chain networks.

Technology exchange involves more than the transfer of knowledge from university to industry. It also involves the transfer of technical manpower to industry as well. As universities work on cutting-edge research, not only do faculty members get engaged in projects, their students also do. University research teams often involve students as much as faculty members and staff. Successful government university research policies not only require that students be integrated into university research as a way to grow their technical skills but also serve as a route to supply the next generation of skilled manpower required by industry and academe. Participation in research helps students develop critical thinking abilities and skills to pursue independent studies. Research also exposes students to the use of technology in solving today's complex problems. Furthermore, knowledge gained through research is integrated into courses to which many students are exposed. The knowledge gained through direct participation in research or classroom instructions eventually find their ways to industry as the students graduate and find employment. Armed with the new skills, these graduates enter the workforce to become the next generation of innovators and technology leaders that industry requires to advance. In this respect, support for research, and consequently, university–industry partnership, is not only a way to grow the knowledge base in the continent but also a way to train skilled manpower to run future high technology business enterprises.

Government support for university–industry partnerships also directly helps ameliorate the research infrastructure problems faced by universities and industry in Africa. Modern research is highly dependent on the use of technology to collect, analyze, and interpret data and results. Due to budget constraints, sheer negligence, and corruption, many Sub-Saharan African universities have been underequipped and underfunded to take their rightful places in the academy. Implementation of policies that support university–industry partnership to promote technology exchange is one avenue to simultaneously accomplish both the task of technology exchange and providing research infrastructure for the universities. Well-thought out university–industry partnership policies on technology exchange would allow for researchers to include, when necessary, requests for infrastructure funding or research equipment acquisition to carry out their proposed research. Sometimes, the research proposal could simply focus on infrastructure or equipment funding to modernize part or an entire research laboratory. While the equipment or laboratory is being used for the proposed research, it is also simultaneously being made available for instructional use to educate students. Furthermore, upon the completion of the research, the acquired equipment becomes part of the research asset of the university to use on other follow-up research programs. Thus, by funding a research program under the university–industry partnership policy, a university automatically acquires modern research equipment to train students and support other related research activities.

The infrastructure acquired as part of a university–industry partnership program is accessible to the partner companies when the research effort is ongoing. This is a sure way for a company to have access to modern equipment they would otherwise not be privileged to because of cost. Many Sub-Saharan African manufacturing enterprises lack the capital to invest in research infrastructure development. Government support for technology exchange programs provides pathways for industry to have access to such infrastructure.

The continual production of quality graduates facilitated by the technology exchange policies of governments will, over time, erase any technical power deficit the region experiences. The sustained production of quality graduates will provide industry access to technical manpower they need to address the problem of poor product quality and global competitiveness. Skilled manpower production should be seen as a priority item for the region in order to address factors that affect high manufacturing cost, low product quality, high labor cost, and low product variety that fail to meet customers' expectations. Therefore, technical manpower production should be attached as a rider to any university–industry partnership policy and with emphasis on student education and training.

University access to real-world problems

In the past, most universities acted as ivory towers. Universities used to isolate themselves from real-world problems and were content with working on problems that stimulate intellectual curiosity with little or no bearings to reality. Graduates enter the job markets with minimal practical knowledge and marketable skills that prepare them for industry work. Today, university–industry partnerships are changing the relationships between universities and industry. Through these relationships, faculty members and students are gaining access and working on problems that are of interest to industry and society in general. This is accomplished through the increased interaction between industry and universities. The interactions enable industry to share problems they encounter in their operations with universities. The variety of problems that industry makes available to universities enables universities to train students and familiarize them with the range of problems they would likely encounter upon entering the workforce. This interaction not only benefits industry, it also benefits universities, faculty, and students alike. In this respect, university–industry relationships are win–win situations for all participating parties. In some cases, the flow of knowledge is reverse. Not all university–industry partnerships involve knowledge transfer from academe to industry. In some cases, the flow of knowledge is from industry to university. In situations where the transfer of knowledge is from industry to university, universities benefit from the collaborations by having access to and work with industry experts with specialized knowledge that they would otherwise not be privileged to in the absence of collaboration. Thus, university–industry partnerships offer tremendous opportunities in which the parties involved share knowledge that may otherwise not be available to one or more of the parties.

Role of university research centers in technology exchange

A university research center is a facility or group of facilities that house a group of researchers from one or more academic disciplines and companies that collaborate on a research effort that is focused on a class of related problems that fall together into a broad research and educational theme. However, the breadth of the problems is such that they cannot be meaningfully be addressed by one faculty. An interdisciplinary approach is essential to address the problems that are the focus of a research center. Successful centers would often have multiple industry partners, faculty, and students that interact freely to execute the research mission of the center. It may also draw members from across multiple colleges on a campus and even multiple universities. In the industrialized economies, governments take active roles in sponsoring research centers that coalesce around research problems that address areas of national needs at university campuses.

Centers are fertile grounds for university researchers to delve deeper into key science and technology problems of national interest at reasonable cost using faculty members. The faculty members, who are already on university payrolls, cost less overall to undertake the research because only part of their time needs to be charged to the center's budget. Student assistants assist the faculty in their research at costs that are considerably lower than the market rate for hiring equivalent full time persons with similar qualifications from the open market. Participating student researchers often work on some specific aspects of the center's research problems and may use their work at the center to satisfy the research requirements in their academic programs.

Industry members play very important role in center operations. Normally, they will make contributions or pay some annual fees for center membership. The fees help to supplement the research funding received from governments. In most cases, total contributions and fees in successful centers would exceed the annual award from government. Industry members have the privilege of having early access to research results generated by the center. These results can be used by industry partners to improve their operations. Industry partners also serve as sources of research problems to the center as long as the problem is within the scope of the center's mission. Industry partners also serve on center's industry advisory board to guide and advise on the center's research direction and priorities. The tight linkage between industry and academia through the work of centers allows for knowledge and technology to flow freely between the two sectors. Government, industry, and universities all benefit in the interaction. Universities use the contributions from industry and the center to strengthen their academic programs, train, and produce students in areas of the economy that is in need of technical manpower. Centers also generate expertise in specific fields to build up the national technical workforce capacity. Furthermore, participation of faculty in a center's work helps them grow their expertise in specific fields and increase their publication rate, and allows them to gain national and international visibility.

Industry partners benefit by having their research problems addressed by a competent team of faculty at a minimal cost because the bulk of the cost is borne by the government in the form of award to the center. Industry members have priority to access the research results. Finally, the interaction at the center provides industry members early access and the opportunity to hire the students trained and participated on the center's research programs. Through their participation in the work of the center, the students will acquire the skills industry needs by the time they graduate. This increases their marketability and career options. At the time of their hire, these students would already be familiar with their employer's problems and therefore do not have the need for extensive industry training to assimilate in their employer's work environment.

The benefits to the government or public are obvious. With increase in industrial productivity resulting from the applications of the research results, the country's export earning is improved through the sale of manufactured products. Improved manufacturing export earning automatically improves the government's tax earnings. Furthermore, the country's pool of technical workforce is broadened by the sustained addition of new graduates each year. A skilled technical workforce increases a country's global competitiveness to attract foreign investment into the industrial sector. Increased investment and productivity create jobs which ultimately improves the economic standing and welfare of the citizens.

Therefore, government policies that support university-based research centers are quick ways to transfer technology between the academic and industrial sectors while simultaneously broadening the technical workforce to boost a country's economic competitiveness.

Technology exchange approaches

Technology exchanges between universities and industry can take one of several forms (Motohashi 2005) that differ significantly in formality and complexity. They range from simple approaches that involve students undertaking industrial projects to satisfy class requirements to more complex arrangements that require the setup of physical areas on campuses in which multiple companies are allocated space to pursue their individual research in collaboration with universities. In particular, the most common forms of technology exchanges are as illustrated graphically in Figure 10.1. They include the following:

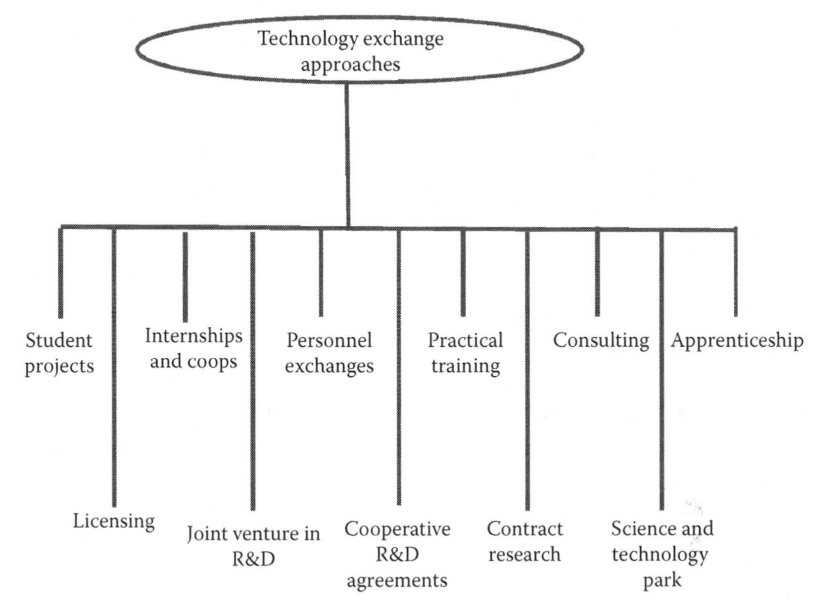

Figure 10.1 Technology exchange mechanisms.

1. Sponsorship of student projects
2. Internships and cooperative education programs
3. Apprenticeships
4. Personnel exchanges
5. Practical training
6. Consulting
7. Joint venture in research and development
8. Cooperative Research and Development Agreement (CRADA)
9. Licensing of intellectual properties
10. Contract research
11. Science and technology parks

Sponsorship of student projects

Sponsorship of student projects is one of the simplest forms of technology exchange that is available. It is also the least expensive. It generally does not require the participation and sponsorship of a government entity. It is a relationship between a company and a university in an informal way in which a student or group of students completing a course requirement is provided access to a problem of interest to a company to execute on behalf of the company. The arrangement for such a relationship may be initiated and negotiated between the company and the students themselves or between the company and the course instructor on behalf of the students. Projects undertaken under this model may last a few weeks to a semester but no longer than a year. They generally involve very minimal cost, which is paid by the company to subsidize the students' effort. In this model, students benefit by applying their knowledge to solve some practical problems and satisfying class requirement, while the company gains from the knowledge of the students to have its problem solved at minimal cost.

Internship and cooperative education

An internship and cooperative education program is a relationship between a university and a company that requires a formal agreement between the two parties. Under this arrangement, a student spends part of his or her time during an academic year at a company site doing some work that is related to his or her curriculum at the university. The intent is for the students to intermix their education with periods of practical training at company sites and on campus. Internship and cooperative education programs may be paid or unpaid. In unpaid internship/cooperative education arrangements, there is no financial obligation on the company. Participating students do so at their own expenses. Paid internships/cooperative education on the other hand requires companies to pay some stipend or salary to the students to cover their expenses and in some cases part of their educational costs. Paid internships/cooperative education is often preferred because they demonstrate the employer's level of commitment and seriousness to the cause of the relationship. It is also an appreciation on the part of the company for the work done by the students. Through this kind of partnership, students gain valuable practical experiences from industry while pursuing their education. Similarly employers gain from the knowledge of the students as they apply that knowledge to the company's problems. Furthermore, a company is able to add additional manpower to its workforce at minimal cost. Students often would return to campus with a much better appreciation and richer understanding of their field of study. The program also helps them to see greater linkage and relevance of the courses they are required to take to the problems that they would likely face in industry upon entering the workforce. A successful internship/cooperative education experience may earn the student an employment with the company upon graduation.

Apprenticeships

Apprenticeship is a process that combines academic training with industry-sponsored research project that is carried out under the supervision of an academic adviser in close contact with a manager that directs the work at the company. The research is conducted both at the company site and on campus. Under this process, a student learns the theoretical aspects of his chosen field at school and undertakes an industry-based research that requires the application of the knowledge so acquired to complete the research. In doing so, the student is directly transferring knowledge acquired in one sector to solve problem in another. The supervisor, working collaboratively with the industry sponsor and keeping the spirit of rigor associated with academic research, ensures that the research quality meets the university expectation for the degree requirement being completed. Academic apprenticeship is most common at the graduate education levels where students are required to complete applied research projects.

Personnel exchange

Personnel exchanges involve temporal transfer of personnel between industry and academia for the purpose of exchanging knowledge that would benefit the partners. The transfer could be from academia to industry in which a faculty member may spend his or her sabbatical at a company site to work on some novel problems of mutual interest or learn new problem-solving techniques or learn the use of new equipment. Possible sites might be company labs, government labs, other universities, or a manufacturing facility that specializes in the production of specialty products involving new high-tech processes. The knowledge gained during the experience is used to strengthen the faculty member's research and educational program on campus. The reverse is also true. In this case, it is a company that sends its employee to a university campus to work with a faculty or team of faculty on joint research projects or learn new processes that would benefit the company. While at the university, the company employee may also be involved in teaching classes in which his or her company has known expertise.

Practical training

Practical training is not new in academic settings. They are extensively used in teacher education programs and in science-based disciplines such as medicine, dentistry, and nursing. Practical training where applicable are considered as part of the curricula. Practical training requires students to complete some field work to apply their academic knowledge and demonstrate the mastery of their profession. It is a collaborative effort between a university and industry to train the next generation of leaders in their respective professions. It benefits academic programs by facilitating the completion of their students' training through the use of industry resources and facilities, and it benefits industry by boosting its staffing level at minimal cost. Furthermore, practical training also provides companies early access to potential future employees that the company can hire to build up its workforce. Most importantly, it serves as a bridge between school and the real world before students complete their training and therefore helps to ease them gradually and seamlessly into the workforce.

Consulting

As a vehicle for knowledge transfer, academic consulting involves the interaction of a researcher or knowledge expert with an industrial user to address a problem of concern to the user. The knowledge expert works on the said problem with some agreed upon goals and deliverables. The relationship requires the identification or definition of what the problem is, what the deliverables are, a timeline for reports, and the cost for the service. From an academic perspective, a faculty can work alone or in collaboration with other experts. The rational is that a research active faculty is in the frontline of new knowledge and that this knowledge is required by industry. Through consulting, the knowledge can be transferred to industry users to innovate their processes. Consulting can be between a faculty member and a company without university involvement or between a university lab and a company. In either case, a contract is required to specify the terms of the relationship. Consulting contracts are usually of short-term durations.

In some cases, the flow of knowledge may be from industry to university. In the United States, for example, the recruitment of industry experts to serve on the industrial advisory boards of academic departments, research programs, centers, and colleges is very common. These advisory board members provide valuable advice and insights that are used by universities to strengthen their curricula or reposition their programs. In fact, the formation of such industrial advisory boards has become the norm rather the exception for science and technology-related academic programs.

Joint venture in research and development

A joint venture is a partnership between a university or consortium of universities and a company or consortium of companies to pursue a joint research and development project and share the cost and profit associated with it. The cost and profit share between the parties may be unequal. Under this model of technology exchange, both the university and the company may contribute the technical personnel required for the project to ensure that the best minds are deployed on it. A formal agreement is required to specify the terms of the cost and profit share and for how long the venture will be in force. A joint venture partnership is focused on a single specific goal for which the parties leverage their resources to undertake the project. Both parties benefit from a joint venture because it spreads the risks. It makes it possible for small companies to participate or undertake research and development projects by reducing their initial expected financial outlay due to cost share. The pooling of technical manpower and financial resources from both parties strengthens and benefits both sides and makes it possible to pursue a project which either party on his or her own will be incapable to undertake. Also worth noting is the fact that the sharing of risks makes undertaking high-risk projects attractive to pursue.

Cooperative research and development agreement

CRADA is a legal relationship between a university laboratory and an industrial team to leverage their resources to pursue research in an area of mutual benefit to the parties that is consistent with the mission of the laboratory (Rogers et al. 1998). In a broader situation, a government entity such as a government lab may also be one of the partners. Under CRADA, the parties share resources to address the problem of interest. A university may contribute personnel, facilities, and other resources, while industry and government members may contribute personnel, facilities, funds, services, equipment, and other resources to undertake the research. In the United States, CRADA is an agreement between a federal laboratory and a nonfederal entity such as a university, industry group, or both. Based on the terms of the agreement, the government may grant the collaborating parties patent licenses on any intellectual properties generated from the research and the right to commercialize the discoveries. The terms of a CRADA will make provisions for resource contributions and obligations of each member, benefit sharing, intellectual property rights, work plan, protection of member proprietary information, liabilities, and other elements that are deemed pertinent. A key benefit of CRADA is the significant reduction in time between laboratory discoveries and the commercialization of the discoveries.

Licensing of intellectual property

Under some legally binding terms of agreement, an owner such as a university (licensor) of an intellectual property may choose to transfer the right to a third party (licensee) such as a company to exploit and use the intellectual property for commercial purposes. Such transfer of right, referred to as licensing, is a very quick and effective way to shorten the time between the discovery of new knowledge and the application of the knowledge to develop and commercialize a new process, product, or service. A license can be exclusive, nonexclusive, coexclusive, or sole. A license is considered exclusive if other than the named licensee, no other third party, including the licensor can exploit the intellectual property for commercial purposes.

A nonexclusive license allows the licensor to retain the right to grant licenses to other people beyond the named licensee to exploit and develop the intellectual property for commercial purposes. In a nonexclusive license, the licensor also retains the right to exploit the intellectual property. A coexclusive license is one in which the licensor can grant transfer rights to multiple licensees but the number of licensees cannot exceed a certain size or value. A sole license is an exclusive license except that the licensor also retains the right to exploit the intellectual property himself. In each case, a licensee may be required to present a plan for commercializing the intellectual property. As one will observe, licensing is an effective way to shorten the time between discovery and the commercialization of the discovery in form of products, services, or processes.

Contract research

A contract research takes place when a company enters into a contract or agreement with a university to undertake a research and development project on its behalf for a fee agreed upon by both parties. The university provides the technical knowledge, personnel, and the facilities to conduct the research. There are clear deliverables specified in the contract as well as the duration. Additionally, other obligations of the company such as the supply of data or access to company proprietary information by the university will also be spelt out if needed. Because contract research is driven by a company's immediate needs, the results are usually put into practice within a short period of time.

Science and technology parks

These are business incubation centers normally located at or close to a university or group of universities with the primary purpose of engaging in scientific research and development projects that are intended to stimulate entrepreneurial, knowledge and technology-based enterprises within a geographic area. A science and technology park will typically house multiple small high-technology companies at different stages in their development. These companies receive government assistance in their early stages. The researchers are personnel provided by the companies themselves and those from the university or universities. Individual companies provide the funding required by their project. Science and technology parks provide infrastructure, rental space, and support services to the resident companies at below market rates. The motivation for science and technology parks is to stimulate economic development by providing nurturing environment to the resident companies at their initial startup stages. By co-locating with a university or universities, technology exchange naturally takes place.

Linkage between technology exchange and policy on science, technology, engineering, and mathematics education

From the discussions in the previous sections, it is clear that there is a clear relationship between technology exchange and the availability of the technical manpower required to carry out the intended research. A country can only reposition itself in the technology market if it has the educated professionals with the right technical skills to undertake the research and development activities. Technology exchange thrives in societies with a pool of technical manpower with research capability to draw from. Many African countries lack the pool of scientists and engineers necessary to carry out the required and essential transformative research to build dynamic economies. While many students graduate each year from universities across Africa, most lack the relevant academic preparation and training to engage in scientific research. Most would graduate without ever having access to well-equipped science and engineering laboratories at their universities even though their educational training or academic majors demand such exposure. This is because the institutions producing the graduates lack the infrastructure and resources to support experiential education in which classroom instructions are supplemented with proper laboratory exercises appropriate for the demands of today's science and technology education. Without experiential learning, the graduates lack the confidence needed to pursue serious research or embark on entrepreneurial activities that are based on science and technology.

Compounding the lack of resources is the lack of up-to-date curricula that reflect the state of science and engineering education in the twenty-first century. Without the availability of the proper infrastructure and access to well-equipped libraries, it is difficult to implement curricula that are anchored on emerging disciplines that tend to be fast changing, dynamic, and equipment intensive. It is in these emerging disciplines such as information technology, nanotechnology/nanoscience, environmental science, energy, and the biological and biomedical sciences that are at the forefront of research that are driving economies today in the rest of the world. The high cost of building the research infrastructure is denying African scientists and engineers participation and contributions in the emerging research areas. With near complete lack of participation in critical research, the universities cannot produce the stream of graduates needed to fuel the pipeline of a technically literate workforce.

For a culture of technology transfer to take place, there has to be a regular stream of scientists and engineers with the right education coming out from the universities annually. Equipped with the proper education, these graduates will be the ones to push the frontiers of innovation and development necessary to fuel the economy and lift Africa to a respectable economic position relative to other regions of the world. The graduates will also be the ones to staff the research laboratories and projects that often form the basis for collaboration between universities and industry to advance innovation and creativity.

Therefore, to build a culture that values technology exchange and economic growth based on a more engaging manufacturing sector, investment must be made in the educational sector to design relevant curricula supported by appropriate infrastructure to produce graduates properly trained and competitive in science, technology, engineering, and mathematics (STEM). To continue the current state of denying African students, the proper STEM education they need due to improper government educational policies is a guarantee that the region will continue to lack meaningful participation in the twenty-first century knowledge economy. This also ensures that the economy of the region will continue to stagnate. Consequently, the lack of participation in manufacturing export will guarantee the current state of affairs, which in the long run spells disaster and instability in the continent.

Conclusion

The low manufacturing export that characterizes the African countries should be a serious concern for governments and leaders in the region. At its current stage in economic development, manufacturing appears to offer the most potential and the most appropriate sector that can lift Africa up from its last place economic position among the major regions of the world. Partnerships between universities and the manufacturing sector as advocated for in this chapter offer one avenue to pursue economic development. However, it alone without other supporting policies will not produce the kinds of results experienced in other regions. Other supporting policies on education, transportation, infrastructure, corruption, taxes, intellectual property protection, workforce development, business climate, and strong legal systems have to be put in place. In fact, global economic competitiveness requires continual reviews of systems and policies to ensure the best practices are in place to support the overall national goals. Commitment to this kind of continuous system review process and sustained holistic approach to economic development can greatly be improved in many countries of the region. This notwithstanding, the issue of university–industry partnerships as an effective instrument for technology exchanges is a necessity for the region as part of a comprehensive set of policies required to grow the region's technical manpower base and economic development.

References

Council on Competitiveness, *Endless Frontier, Limited Resources: U.S. R&D Policy for Competitiveness*, Council on Competitiveness, Washington, DC, 1996.

Debackere, K., R. Veugelers, The role of academic technology transfer organizations in improving industry science links, *Research Policy*, Vol. 34, pp. 321–342, 2005.

KPMG, *Sector Report on Manufacturing in Africa*, KPMG, Africa, 2014.

Lee, J., H. N. Win, Technology transfer between university research centers and industry in Singapore, *Technovation*, Vol. 24, pp. 433–442, 2004.

Motohashi, K., University–industry collaborations in Japan: The role of new technology-based firms in transforming the National Innovation System, *Research Policy*, Vol. 34, pp. 583–594, 2005.

Mowery, D., R. Nelson, B. Sampat, A. Ziedonis, The growth of patenting and licensing by U.S. universities: An assessment of the effects of the Bayh-Dole act of 1980, *Research Policy*, Vol. 30, No. 1, pp. 99–119, 2001.

Nelson, R., Observations on the Post-Bayh–Dole rise in patenting at American universities, *The Journal of Technology Transfer*, Vol. 26, pp. 13–19, 2001.

Rogers, E. M., E. G. Carayannis, K. Kurihara, M. M. Allbritton, Cooperative Research and Development Agreements (CRADAs) as technology transfer mechanisms, *Research and Development Management*, Vol. 28, No. 2, pp. 79–88, 1998.

U.S. Government Accounting Office (GAO) Report to Congressional Committees, *Technology Transfer, Administration of the Bayh-Dole Act by Research Universities*, Washington, DC, May 7, 1998.

chapter eleven

Industry–university collaboration[*]

Pressures from the knowledge-based economy and university–industry collaboration are transforming the traditional role of universities. This chapter explores the use of a university–industry collaboration approach and presents an action plan for establishing a manufacturing research center at The University of the West Indies (UWI) in Trinidad and Tobago. Under the auspices of the Department of Mechanical and Manufacturing Engineering of the University, the center is to foster the university–industry collaboration and bring synergy of research outputs within the department and across the faculty of engineering and other faculties of the university.

Introduction

Business conditions in the twenty-first century have become increasingly complex and dynamic. New approaches to positioning, aligning, process, and technology continually suggest the need for radical and complicated change. Manufacturers are confronted by the inherent complexities and face new standards for globalization, consortium development, and supply chain management (DTI 2000).

Within the rapidly changing world of higher education, links between universities and industry are assuming ever-increasing importance (Jones and Harris 1995). The range of business and education links is varied and includes contract research and consultancy, short-course programs, and the introduction of award-bearing courses specifically designed and/or run for client companies, and in more collaborative sense, the establishment of business development and research centers within universities. University–industry collaboration has the potential to encourage higher level learning. It provides the possibilities not only of learning about new technologies, but of learning about methods of creating future technologies and of the ways those technologies might affect the existing business (Dodgson 1993; Barañano 1999). The collaborators

[*] Adapted with permission from Kit Fai Pun, Ruel Ellis, and Winston G. Lewis, A university-industry collaboration agenda for establishing a manufacturing research center at UWI, *Proceedings of the 20th International Conference on CAD/CAD, Robotics and Factories of the Future*, San Cristobal, Venezuela, pp. 296–303, July 2004.

could share new and more effective ways of doing things and sharpen their competitiveness.

This chapter discusses the need of establishment and operation of a manufacturing research center at the UWI in Trinidad and Tobago. It explores the use of an university–industry collaboration approach and presents an action plan for establishing the center.

University–industry collaboration

Dodgson (1993) defines collaboration as any activity where two or more partners contribute differential resources and know-how to agreed complementary aims. This could be created and promoted by (1) collaborative research programs or consortia, (2) joint ventures and strategic alliances, and (3) shared R&D and production contracts. Many recent studies (e.g., Cyert and Goodman 1997; Keithley and Redman 1997; Barañano 1999; Etzkowitz et al. 2000; Logar et al. 2001; Neumann and Banghart 2001; Hagen 2002) have investigated the need and practical implications for forging closer links, alliances, and partnerships between industry and academia. Others (e.g., see MacPherson 1997; Nicholson 2000) have addressed the approaches and shared the experience of developing research and technology centers or the like.

A number of problems could arise because of the different nature of universities and industry partners. Two major divergences can be found between both kinds of entities. On the one hand, while industry looks for marketable outputs and is not very concerned with the production of knowledge, universities place great emphasis on the creation of advanced, leading-edge knowledge, regardless of its commercial potential. On the other hand, while universities regard the duration of their contracts with industry extremely short, firms consider them as far too long (Barañano 1999).

Nevertheless, cooperation between firms and universities opens a possible door to synergies. On the university side, cooperation is a way of both overcoming shortfalls in the traditional sources of funds and maintaining these institutions at the desired teaching and research standards. On the industry side, cooperation provides a solution to the problems of dealing single-handedly with the challenge of innovation, as well as a way of gaining privileged access to the best students in order to recruit them (Barañano 1999). The university–industry collaboration could (1) foster innovation and research and development (R&D) activities, (2) facilitate technology transfer and the development of viable businesses in the community, and (3) provide continuing professional development and to keep up people to maintain their skills and effectiveness that meet the needs of a nation.

Needs of university–industry collaboration: A Trinidad case

Increased competition, sophisticated customer demands, and rapid advances in technology drive the industry and the society at large. In developing countries, R&D must be focused and self-sustaining. This must bring available and indigenous technologies to bear on the socioeconomic issues facing the society. In this context, promoting a strong university–industry partnerships/alliance and closer collaboration could provide a platform and a stimulator for R&D that could help identify the driving forces of change and coping with the resultant changes (Hagen 2002).

In Trinidad and Tobago, the energy sectors and related industries have been dominating the economic development of the country. The government has realized the importance of achieving a balanced and sustainable growth. The growing interdependence of manufacturing and services reinforces the need for an innovative, creative, and forward-looking manufacturing capacity in the future.

In 2002, the manufacturing sector accounted for about 11% of total employment (UWI 2003). For the past 4 years, employment in the manufacturing sector has grown at the same rate as the country's gross domestic product. The four largest manufacturing groups by core occupation are (1) food and food process; (2) fabricated metals; (3) wood products; and (4) chemical, rubber, and plastic. The growth of the Trinidad and Tobago manufacturing sector, even at the low-end estimate (1.5% growth per year), supports the need of a manufacturing technology/research center (UWI 2003).

Faced with today's ever-increasing competitors and changing industrial environment, there has been a pressing need to initiate closer collaboration among the government, the industry, the universities, and other stakeholders (Etzkowitz et al. 2000). In the case of Trinidad and Tobago, the collaboration would (1) promote redevelopment and re-engineering the manufacturing sector; (2) facilitate and initiate industry-wide cooperation to develop generic solutions, reducing the cost structure and improving the economic efficiency of the manufacturing sector; and (3) ensure that all levels of the workforce have the right skills and training.

Collaborative research center at UWI

In responding to the need for closer university–industry collaborations, the UWI Faculty of Engineering has taken an initiative to establish an Enterprise Research Integration Centre (ERIC) under the Department of the Mechanical and Manufacturing Engineering (Pun 2003). In 2003, a task force was formed to plan for the establishment of the center. A center venue was selected, and the search for start-up equipments, hardware, and software has been undertaken. Some refurnishing works have been started. It is expected that the center could be in full operation by end of 2005.

Mission and functions of the center

The mission of the center is to promote teaching and research in advanced manufacturing technology in the university and to work closely with the industry through consultancy, collaborative projects, and specialized programs to disseminate best practices in Trinidad and Tobago, as well as the Caribbean region.

The main emphasis of the center's activities would be placed on the provision of services including applied and contract research, consultancy, and manpower training. The center has the following three main functions:

1. To promote a wider development of individual and collective engineering projects relying on the more flexible generation; processing and application of strategic information, knowledge, and human-like intelligence for enhancing the overall value; and quality of engineering processes, products, or systems
2. To introduce innovative technology and consultancy services by identifying and assessing the business potential of research projects and technology commercialization
3. To support research activities and teaching programs at UWI as well as to provide specifically designed company courses for industry partners

Establishment action plan of the center

Several forms of university–industry collaboration are being investigated to assure a proper establishment and operations of the center in line with its mission and goals to serve the nation. The collaboration would depend upon the identification and indoctrination of possible industrial organizations, the identification of products and processes that merit academic energies, and focus that is mutually beneficial to both the university and industry partners.

In order to facilitate the establishment of the center, the ERIC task force has formulated an action plan addressing the key areas given as follows:

1. To conduct a feasibility study on examining the SWOT (strengths, weaknesses, opportunities, and threats) of establishing a manufacturing research center at UWI
2. To identify core functions, design facility layout, and determine feasible operational structure for the center
3. To determine niche areas in product design and manufacture that the center can concentrate upon

4. To build its capabilities and collaboration base with other existing centers and institutes in the faculty of the engineering, as well as within the university
5. To strengthen its alliances with industry partners and other research institutes in the local and regional context
6. To evaluate the effectiveness of university–industry collaboration and enhance the collaboration it has with results from pilot projects and case studies

Stages of the establishment

An initial establishment plan for the center has four stages and 12 tasks. These stages are (1) the project initiation and feasibility study, (2) functional analysis and design, (3) infrastructure building and trial operation, and (4) evaluation and operation audit. The 12 tasks are scheduled to complete in 2 years from January 2004 to December 2005 (see Table 11.1). The duration of individual stages are overlapped to ensure their nature of continuity.

Table 11.1 Highlights of an Establishment Plan

Stages	Tasks
I: Project Initiation and Feasibility Study (*January–August 2004: 8 months*)	1. Review from published literature (books, articles, reports, etc.) on different approaches on establishing a research center 2. Elicitation of documents and materials from industry, Internet, and other sources 3. Conduct a SWOT analysis of establishing the center at UWI and identification and indoctrination of possible industrial partners
II: Functional Analysis and Design (*May–December 2004: 8 months*)	1. Analyze the core functions and determine the niches of the center 2. Design the collaboration platform, system architecture, user interface, and facility/layout designs with respect to the structure of the center 3. Acquire equipment and software and create the platform from which relevant research activities and the services would be launched
III: Infrastructure Building and Trial Operation (*November 2004–August 2005: 10 months*)	1. Install equipment installation and develop infrastructure for different functional areas in the center 2. Develop specifications and descriptions for the projects identified for facilitating undergraduate studies and postgraduate research 3. Identify and initiate potential projects (including consultancy)

(*Continued*)

Table 11.1 (Continued) Highlights of an Establishment Plan

Stages	Tasks
IV: Evaluation and Operation Audit (*July–December 2005: 6 months*)	1. Perform selected projects to demonstrate the effectiveness of the university–industry collaboration approach 2. Document the process and identify opportunities available to the center 3. Recommend changes wherever applicable

A proposed center organization

A formal organization of the ERIC Board would administer the operations of the center. The board would comprise of faculty members, academic staff, and representatives from industry. Initially, the center would have three extension arms, namely, the innovative research and development (IRD) division, the manufacturing consultancy and services (MCS) division, and the manpower training and intelligence (MTI) division. The center would expand its scope and functions. Exploring other specialized disciplines, divisions, and subdivisions (e.g., advanced manufacturing, polymer processing) is under investigation. A diagrammatic representation of the initial organizational structure of the center is shown in Figure 11.1. A brief description of individual divisions and their core functions is provided in the subsequent sections.

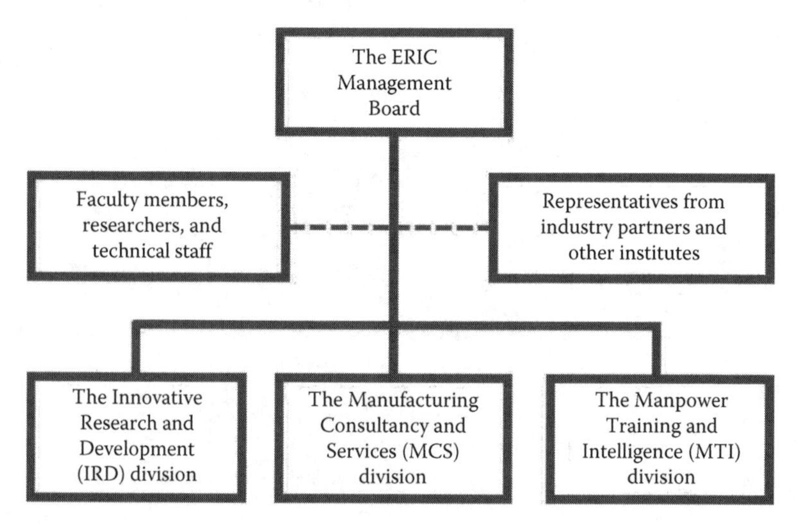

Figure 11.1 The initial organizational structure of ERIC.

Innovative research and development division

This division would be set up as an area of excellence and as a resource base for collaborative work to foster innovation and R&D activities, consistent with the center's mission. With a closer collaboration with industry partners and other research institutes in the university (e.g., the Advanced Mechanical Engineering [ADMECH] Centre, the Engineering Institute) and outside (e.g., the National Institute of Higher Education of Science and Technology and Caribbean Industrial Research Institute, Trinidad and Tobago of Standards), the IRD division would promote the wider development of individual and collective engineering projects.

This division would encourage and support the following major activities:

- Mechanical and nonmechanical design processes
- Automation
- Instrumentation
- Engineering information systems and management
- Rapid prototype and product development
- Knowledge-based virtual enterprise
- E-manufacturing

Manufacturing consultancy and services division

With the support from faculty members and academic staff, the MCS division would assist industry partners to identify and assess the business potential of research projects and technology commercialization. It provides a bundle of consultancy services that facilitate technology transfer and the development of viable businesses in the community. Its core services include the following:

- To promote commercialization of the universities' and industry partners' research and technologies
- To identify and develop business opportunities
- To incubate companies and provide value-added services, including management and professional advisory services, consultancy, and technical support

Manpower training and intelligence division

Working closely with the other two divisions and the Engineering Institute of the Faculty of Engineering, the MTI division would build its intelligence and information bases with focus on manpower resource and training. Its services would include the provision of support on (1) the traditional teaching/training courses offered and (2) the collaboration research activities and projects in manpower development and training within the faculty of engineering and other faculties of the university, as well as industry partners.

The division would also provide specifically designed company programs and courses for industry partners and keep participants in the knowledge they need to maintain their skills and effectiveness to meet their needs. This would support local manufacturers to up-skill or multiskill their engineers, technicians, and operators on site and provides useful training that will have a significant impact back in the workplace.

Conclusion

Working with the industry is a vital part of the university's mission. In this way, the ERIC would provide manufacturing enterprises with access to specialist consultancy, R&D, technology suppliers, networks, and training services. Adopting university–industry collaboration could promote the synergy between industry and academia. The collaboration would facilitate functional integration of the center and enhance the accessibility of its services to a broader community of users (such as organizations or individuals) either for educational or for commercial purpose.

In the academia context, growing benefits could be identified, for example, in terms of company- and industry-based student projects, research and staff development, and the development of new teaching materials and sharing of expertise across organizational boundaries.

In the industry part, the center could provide manufacturing enterprises with access to specialist consultancy, R&D, technology suppliers, networks, and training services. It is expected that the center could work with an extensive range of industrial and academic partners, including other local and overseas institutes.

Looking to the future, university–industry collaboration would become an increasingly common feature in Trinidad and Tobago, as well as in the regional or global context. In this light, the ERIC case provides some interesting perspectives on how such collaboration approach can be managed to the benefits of the involved parties and other stakeholders.

References

Barañano, A. M., Understanding technological collaboration: A literature review on some key issues, *Dablium*, Vol. 1, No. 3, available at: http://members.lycos.co.uk/Dablium/artigo10.htm (accessed January 15, 2004), 1999.

Cyert, R. J., P. S. Goodman, Creating effective university-industry alliances: An organizational learning perspective, *Organizational Dynamics*, Vol. 25, No. 4, p. 45, 1997.

Dodgson, M., *Technological Collaboration in Industry: Strategy, Policy and Internationalization in Innovation*, Routledge, London, 1993.

DTI, *DTI Manufacturing Website—Gateway to the Manufacturing Advisory Service*, available at: http://www.dti.gov.uk/manufacturing/ (accessed January 15, 2004), 2000.

Etzkowitz, H., A. Webster, C. Gebhardt, B. R. C. Terra, The future of the university and the university of the future: Evolution of ivory to entrepreneurial paradigm, *Research Policy*, Vol. 29, No. 2–3, pp. 313–330, 2000.

Hagen, R., Globalization, university transformation and economic regeneration: A UK case study of public/private sector partnership, *International Journal of Public Sector Management*, Vol. 15, No. 3, pp. 204–218, 2002.

Jones, P., I. Harris, Educational and commercial collaboration in higher education: A case study from retailing, *Industrial and Commercial Training*, Vol. 27, No. 1, pp. 4–8, 1995.

Keithley, D., T. Redman, University-industry partnerships in management development: A case study in a 'world-class' company, *The Journal of Management Development*, Vol. 16, No. 3, pp. 154–166, 1997.

Logar, C. M., T. G. Ponzurick, J. R. Spears, K. R. France, Commercializing intellectual property: A university-industry alliance for new product development, *Journal of Product and Brand Management*, Vol. 15, No. 1, pp. 7–11, 2001.

MacPherson, R. J. S., The centre for professional development at the University of Auckland: Towards creating networks of moral obligations, *International Journal of Educational Management*, Vol. 11, No. 6, pp. 260–267, 1997.

Neumann, B. R., S. Banghart, Industry-university 'consulternships': An implementation guide, *The International Journal of Educational Management*, Vol. 15, No. 1, pp. 7–11, 2001.

Nicholson, D., Researching and developing virtual Scotland—A perspective from the Centre for Digital Library Research, *The Electronic Library*, Vol. 18, No. 1, pp. 51–62, 2000.

Pun, K. F., The establishment of an Enterprise Research Integration Centre, *Internal Document*, Department of Mechanical and Manufacturing Engineering, The University of the West Indies, June 2003.

UWI, *Manufacturing and Micro-Enterprise: Summary Draft*, The University of the West Indies, Trinidad and Tobago, 2003.

chapter twelve

Project management for technology transfer

Nothing gets done on time, on cost, and on target without the application of project management (PM). Target in this regard refers to the quality expectations or performance specifications of the end product. This chapter presents an introduction to the tools and techniques of PM.

What is a project?

A project is traditionally defined as a unique one-of-a-kind endeavor with a specific goal that has a definite beginning and a definite end. Project Management Body of Knowledge (PMBOK®) defines a project as a temporary endeavor undertaken to create a unique product, service, or result. Temporary means having a defined beginning and a definite end. The term *unique* implies that the project is different from other projects in terms of characteristics.

What is PM?

The author defines PM as the process of managing, allocating, and timing resources to achieve a given goal in an efficient and expeditious manner.

PMBOK® defines PM as the application of knowledge, skills, tools, and techniques to project activities to achieve project objectives.

The rapid growth of technology in the work has created new challenges for those who plan, organize, control, and execute complex projects. With the diversity of markets globally, project integration is of great concern. Using a consistent body of knowledge can alleviate the potential problems faced in the project environment. Projects related to technology transfer are particularly complex and dynamic, thus necessitating a consistent approach. The use of PM continues to grow rapidly. The need to develop effective management tools increases with the increasing complexity of new technologies and processes. The life cycle of a new product to be introduced into a competitive market is a good example of a complex process that must be managed with integrative PM approaches. The product will encounter management functions as it goes from one stage to the next. PM will be needed throughout the design and production stages of the product. PM will be needed in developing marketing, transportation, and delivery strategies for the product. When the product finally gets to the customer, PM will be needed to integrate its use with those of other products within the customer's organization. The need for a PM approach is established by the fact that a project will always tend to increase in size even if its scope is narrowing. The following four PM rules are applicable to any project environment:

1. *Parkinson's law*: Work expands to fill the available time or space.
2. *Peter's principle*: People rise to the level of their incompetence.
3. *Murphy's law*: Whatever can go wrong will.
4. *Badiru's rule*: The grass is always greener where you most need it to be dead.

An integrated systems PM approach can help diminish the negative implications suggested by these rules through good project planning, organizing, scheduling, and control. The following sections present some elements of the PMBOK®, published by the Project Management Institute.* The nine knowledge areas presented in the PMBOK® are listed below:

1. Integration
   a. Integrative project charter
   b. Project scope statement
   c. PM plan
   d. Project execution management
   e. Change control
2. Scope management
   a. Focused scope statements
   b. Cost/benefits analysis
   c. Project constraints
   d. Work breakdown structure (WBS)
   e. Responsibility breakdown structure
   f. Change control
3. Time management
   a. Schedule planning and control
   b. PERT and Gantt Charts
   c. Critical path method
   d. Network models
   e. Resource loading
   f. Reporting
4. Cost management
   a. Financial analysis
   b. Cost estimating

* http://www.pmi.org/PMBOK-Guide-and-Standards.aspx (accessed April 7, 2015).

   c. Forecasting
   d. Cost control
   e. Cost reporting
5. Quality management
   a. Total quality management
   b. Quality assurance
   c. Quality control
   d. Cost of quality
   e. Quality conformance
6. Human resources management
   a. Leadership skill development
   b. Team building
   c. Motivation
   d. Conflict management
   e. Compensation
   f. Organizational structures
7. Communications
   a. Communication matrix
   b. Communication vehicles
   c. Listening and presenting skills
   d. Communication barriers and facilitators
8. Risk management
   a. Risk identification
   b. Risk analysis
   c. Risk mitigation
   d. Contingency planning
9. Procurement and subcontracts
   a. Material selection
   b. Vendor pre-qualification
   c. Contract types
   d. Contract risk assessment
   e. Contract negotiation
   f. Contract change orders

The above segments of the PMBOK® cover the range of functions associated with any project, particularly complex ones.

PM methodology

A PM methodology (PMM) defines a process that a project team uses in executing a project, from planning through phaseout. People, process, and technology assets (science and engineering) form the basis for implementing organizational goals. Human resources constitute crucial capital that must be recruited, developed, and preserved. Organizational work process must take advantage of the latest tools and techniques such as business process reengineering, continuous process improvement, Lean, Six-Sigma, and systems thinking. The coordinated infrastructure represents the envelope of operations and includes physical structures, energy, leadership, operating culture, and movement of materials. The ability of an organization to leverage science and technology to move up the global value chain requires the softer side of PM in addition to the technical techniques. Another key benefit of applying integrative PM to technology transfer projects involves safety. Technology undertakings are volatile and subject to safety violations through one of the following actions:

1. Systems or individuals who deliberately, knowingly, willfully, or negligently violate embedded safety requirements in science, technology, and engineering projects
2. Systems or individuals who inadvertently, accidentally, or carelessly compromise safety requirements in science, technology, and engineering projects

The above potential avenues for safety violation make safety training, education, practice, safety monitoring, and ethics very essential for technology transfer applications. An integrative approach to PM helps to cover all the possible ways for safety compromise.

What is a program?

A program is defined as a recurring group of interrelated projects managed in a coordinated and synergistic manner to obtain integrated results that are better than what is possible by managing the projects individually. Programs often include elements of collateral work outside the scope of the individual projects. Thus, a program is similar to having a system of systems of projects, whereby an entire enterprise might be affected. While projects have definite end points, programs often have unbounded life spans.

Stakeholders are individuals or organizations whose interests may be positively or negatively impacted by a project. Stakeholders must be identified by the project team for every project. A common deficiency in this requirement is that organization employees are often ignored, neglected, or taken for granted as stakeholders in projects going on in the organization. As the definition of stakeholders clearly suggests, if the interests of the employees can be positively or negatively affected by a project, then the employees must be viewed as stakeholders. All those who have a vested interest in the project are stakeholders and this might include the following:

- Customers
- Project sponsor

- Users
- Associated companies
- Community
- Project manager
- Owner
- Project team members
- Shareholders

PM processes

The major knowledge areas of PM are administered in a structured outline covering five basic clusters listed below:

1. Initiating
2. Planning
3. Executing
4. Monitoring and controlling
5. Closing

The implementation clusters represent five process groups that are followed throughout the project life cycle. Each cluster itself consists of several functions and operational steps listed below:

1. Integration
2. Scope
3. Time
4. Cost
5. Quality
6. Human resources
7. Communication
8. Risk
9. Procurement

When the clusters are overlaid on the nine knowledge areas, we obtain a two-dimensional matrix that spans 43 major process steps as shown in Table 12.1. The monitoring and controlling clusters are usually administered as one lumped process group (monitoring and controlling). In some cases, it may be helpful to separate them to highlight the essential attributes of each cluster of functions over the project life cycle. In practice, the processes and clusters do overlap. Thus, there is no crisp demarcation of when and where one process ends and where another one begins over the project life cycle. In general, project life cycle defines the following:

1. Resources that will be needed in each phase of the project life cycle
2. Specific work to be accomplished in each phase of the project life cycle

Table 12.1 Overlay of Project Management Areas and Implementation Clusters

PM knowledge areas	Project management process clusters				
	Initiating	Planning	Executing	Monitoring and controlling	Closing
Project integration	Develop project charter Develop preliminary project scope	Develop project management plan	Direct and manage project execution	Monitor and control project work Integrated change control	
Scope		Scope planning Scope definition Create WBS		Scope verification Scope control	
Time		Activity definition Activity sequencing Activity resource estimating Activity duration estimating Schedule development		Schedule control	
Cost		Cost estimating Cost budgeting		Cost control	
Quality		Quality planning	Perform quality assurance	Perform quality control	

(*Continued*)

Table 12.1 (Continued) Overlay of Project Management Areas and Implementation Clusters

PM knowledge areas	Project management process clusters				
	Initiating	Planning	Executing	Monitoring and controlling	Closing
Human Resources		Human resource planning	Acquire project team Develop project team	Manage project team	
Communication		Communication planning	Information distribution	Performance reporting Manage stakeholders	
Risk		Risk management planning Risk identification Qualitative risk analysis Quantitative risk analysis Risk response planning		Risk monitoring and control	
Procurement		Plan purchases and acquisitions Plan contracting	Request seller responses Select sellers	Contract administration	Contract closure

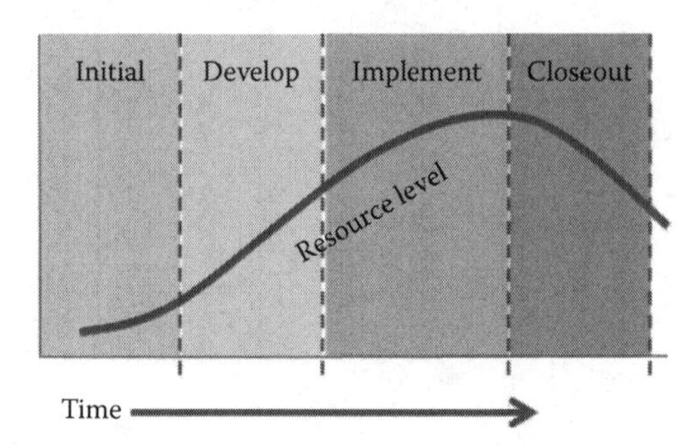

Figure 12.1 Phases in the technology project life cycle.

Figure 12.1 shows the major phases of project life cycle going from the conceptual phase through the closeout phase. It should be noted that project life cycle is distinguished from product life cycle. Project life cycle does not explicitly address operational issues, whereas product life cycle is mostly about operational issues starting from the product's delivery to the end of its useful life. It is noted that the shape of the life cycle curve may be expedited due to rapid developments that often occur in new technology endeavors. For example, for an exploration technology project, the entire life cycle may be shortened, with a very rapid initial phase, even though the conceptualization stage may be very long. Typical characteristics of project life cycle include the following:

1. Cost and staffing requirements are lowest at the beginning of the project and ramp up during the initial and development stages.
2. The probability of successfully completing the project is lowest at the beginning and highest at the end. This is because many unknowns (risks and uncertainties) exist at the beginning of the project. As the project nears its end, there are fewer opportunities for risks and uncertainties.
3. The risks to the project organization (project owner) are lowest at the beginning and highest at the end. This is because not much investment has gone into the project at the beginning, whereas much has been committed by the end of the project. There is a higher sunk cost manifested at the end of the project.
4. The ability of the stakeholders to influence the final project outcome (cost, quality, and schedule) is highest at the beginning and gets progressively lower toward the end of the project. This is intuitive because influence is best exerted at the beginning of an endeavor.
5. Value of scope changes decreases over time during the project life cycle while the cost of scope changes increases over time. The suggestion is to decide and finalize scope as early as possible. If there are to be scope changes, do them as early as possible.

The specific application context will determine the essential elements contained in the life cycle of the endeavor. Life cycles of business entities, products, and projects have their own nuances that must be understood and managed within the prevailing organizational strategic plan. The components of corporate, product, and project life cycles are summarized as follows:

- *Corporate (business) life cycle:*
    - Planning → Needs → Business conceptualization → Realization → Portfolio management
- *Product life cycle:*
    - Feasibility studies → Development → Operations → Product obsolescence
- *Project life cycle:*
    - Initiation → Planning → Execution → Monitoring and Control → Closeout

There is no strict sequence for the application of the knowledge areas to a specific project. The areas represent a hybrid of processes that must be followed in order to achieve a successful project. Thus, some aspects of integration may be found under the knowledge area for communications. In a similar vein, a project may start with the risk management process before proceeding into the integration process. The knowledge areas provide general guidelines. Each project must adapt and tailor the recommended techniques to the specific need and unique circumstances of the project.

Specific strategic, operational, and tactical goals and objectives are embedded within each step in the loop. For example, *initiating* may consist of project conceptualization and description. Part of *executing* may include resource allocation and scheduling. *Monitoring* may involve project tracking, data collection, and parameter measurement. *Controlling* implies taking corrective action based on the items that are monitored and evaluated. *Closing* involves phasing out or terminating a project. Closing does not necessarily mean a death sentence for a project, as the end of one project may be used as the stepping stone to the next series of endeavors.

Factors of technology success or failure

There are several factors that impinge on the success or failure of a project. In technology projects, factors that enhance project success include the following:

- Well-defined scope
- Communication among project team members
- Cooperation of project teams

- Coordination of project efforts
- Proactive management support
- Measurable metrics of project performance
- Identifiable points of accountability
- Realistic time, cost, and requirements

When projects fail, it is often due to a combination of the following factors related to project requirements:

- Requirements are incomplete
- Poor definition of project objectives
- Poor definition of scope and premature acceptance
- Requirements are unrealistic
- Requirements are ambiguous
- Requirements are inconsistent
- Changes in requirements are unbudgeted
- Poor management support
- Lack of alignment of project objectives with organizational objectives
- Poor communication
- Lack of cooperation
- Deficient coordination of project efforts

Work breakdown structure

WBS represents the foundation over which a project is developed and managed. WBS refers to the itemization of a project for planning, scheduling, and control purposes. WBS defines the scope of the project. In the project implementation template, WBS is developed within the scope knowledge area under the planning cluster. The WBS diagram presents the inherent components of a project in a structured block diagram or interrelationship flow chart. WBS shows the relative hierarchies of parts (phases, segments, milestone, etc.) of the project. The purpose of constructing a WBS is to analyze the elemental components of the project in detail. If a project is properly designed through the application of WBS at the project planning stage, it becomes easier to estimate cost and time requirements of the project. Project control is also enhanced by the ability to identify how components of the project link together. Tasks that are contained in the WBS collectively describe the overall project goal. Overall project planning and control can be improved by using a WBS approach. A large project may be broken down into smaller sub-projects that may, in turn, be systematically broken down into task groups. Thus, WBS permits the implementation of a *divide and conquer* concept for project control.

Individual components in a WBS are referred to as WBS elements, and the hierarchy of each is designated by a level identifier. Elements at the same level of subdivision are said to be of the same WBS level. Descending levels provide increasingly detailed definition of project tasks. The complexity of a project and the degree of control desired determine the number of levels in the WBS. Each component is successively broken down into smaller details at lower levels. The process may continue until specific project activities (WBS elements) are reached. In effect, the structure of the WBS looks very much like an organizational chart. But it should be emphasized that WBS is not an organization chart. The basic approach for preparing a WBS is given subsequently.

Level 1 WBS

This contains only the final goal of the project. This item should be identifiable directly as an organizational budget item.

Level 2 WBS

This level contains the major subsections of the project. These subsections are usually identified by their contiguous location or by their related purposes.

Level 3 WBS

Level 3 of the WBS structure contains definable components of the level 2 subsections. In technical terms, this may be referred to as the finite element level of the project.

Subsequent levels of WBS are constructed in more specific details depending on the span of control desired. If a complete WBS becomes too crowded, separate WBS layouts may be drawn for the level 2 components. A statement of work (SOW) or WBS summary should accompany the WBS. The SOW is a narrative of the work to be done. It should include the objectives of the work, its scope, resource requirements, tentative due date, and feasibility statements. A good analysis of the WBS structure will make it easier to perform scope monitoring, scope verification, and control project work later on in the project.

The matrix organization structure

The matrix organization structure is a frequently used organization structure in business and industry. It is particularly applicable to technology management scenarios where inter-departmental interfaces are essential. It is used where there is multiple managerial accountability and responsibility for a project. It combines the advantages of the traditional structure and the product organization structure. The hybrid configuration of the

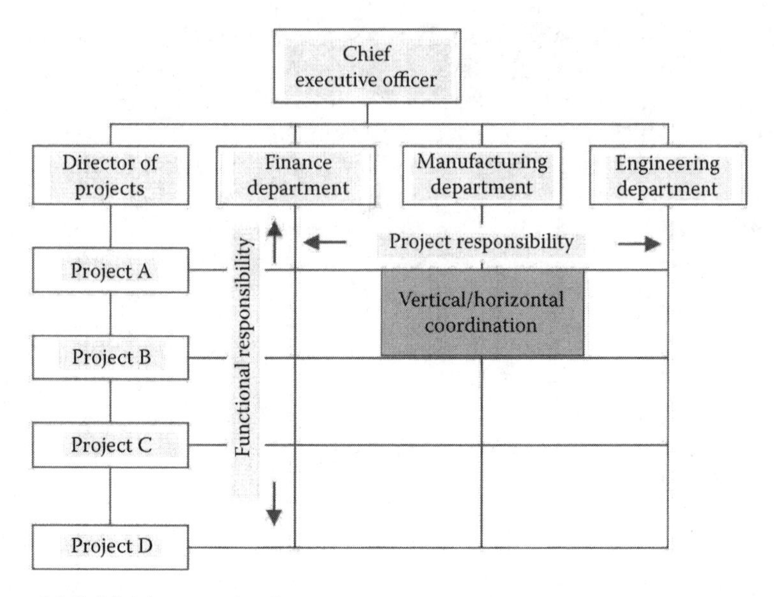

Figure 12.2 Matrix organization structure.

matrix structure facilitates maximum resource utilization and increased performance within time, cost, and performance constraints. There are usually two chains of command involving both horizontal and vertical reporting lines. The horizontal line deals with the functional line of responsibility, while the vertical line deals with the project line of responsibility. An example of a matrix structure is shown in Figure 12.2. The personnel along each vertical line of reporting cross over horizontally to work on the *matrixed* project. The matrix structure is said to be *strong* if it is more closely aligned with projectized organization structure, and it is said to be a weak matrix structure if it is more closely aligned to a functional structure. A balanced matrix structure blends projectized and functional structures equally.

Advantages of matrix organization include the following:

- Good team interaction.
- Consolidation of objectives.
- Multilateral flow of information.
- Lateral mobility for job advancement.
- Individuals have an opportunity to work on a variety of projects.
- Efficient sharing and utilization of resources.
- Reduced project cost due to sharing of personnel.
- Continuity of functions after project completion.
- Stimulating interactions with other functional teams.
- Functional lines rally to support the project efforts.

- Each person has a *home* office after project completion.
- Company knowledge base is equally available to all projects.

Some of the disadvantages of matrix organization are summarized as follows:

- Matrix response time may be slow for fast-paced projects.
- Each project organization operates independently.
- Overhead cost due to additional lines of command.
- Potential conflict of project priorities.
- Problems inherent in having multiple bosses.
- Complexity of the structure.

Traditionally, industrial projects are conducted in serial functional implementations such as R&D, engineering, manufacturing, and marketing. At each stage, unique specifications and work patterns may be used without consulting the preceding and succeeding phases. The consequence is that the end product may not possess the original intended characteristics. For example, the first project in the series might involve the production of one component while the subsequent projects might involve the production of other components. The composite product may not achieve the desired performance because the components were not designed and produced from a unified point of view. The major appeal of matrix organization is that it attempts to provide synergy within groups in an organization.

Elements of a project plan

A project plan represents the road map for executing a project. It contains the outline of the series actions needed to accomplish the project goal. Project planning determines how to initiate a project and execute its objectives. It may be a simple statement of a project goal or it may be a detailed account of procedures to be followed during the project life cycle. In a project plan, all roles and responsibilities must be clearly defined. A project plan is not a bar chart or Gantt chart. The project manager must be versatile enough to have knowledge of most of the components of a project plan. The usual components of a detailed project plan include the following:

- Scope planning
- Scope definition
- WBS
- Activity definition
- Activity sequencing
- Activity resource estimating

- Activity duration estimating
- Schedule development
- Cost estimating
- Cost budgeting
- Quality plan
- Human resource plan
- Communications plan
- Risk management plan
- Risk identification
- Qualitative and quantitative risk analysis
- Risk response planning
- Contingencies
- Purchase plan
- Acquisition plan
- Contracting plan

Integrated systems approach to technology projects

PM tools for technology transfer projects can be classified into the following three major categories described below:

1. *Qualitative tools*: These are the managerial tools that aid in the interpersonal and organizational processes required for PM.
2. *Quantitative tools*: These are analytical techniques that aid in the computational aspects of PM.
3. *Computer tools*: These are computer software and hardware tools that simplify the process of planning, organizing, scheduling, and controlling a project. Software tools can help in with both the qualitative and quantitative analyses needed for PM.

Managing technology requirements

Henry Ford once said his Model T automobile customers could have a car painted any color that he wants so long as it is black. But the fact is that Ford initially offered three colors: green, bright red, and green from 1908 through 1914. But when his production technology advanced to the stage of mass production on moving assembly line, the new process required a fast-drying paint, and only one particular black paint pigment met the requirements. Thus, as a result of the emergence of fast-moving mass production lines, Ford was forced to limit color options to black only. This led to the need for the famous quote. The black-only era spanned the period from 1914 through 1925, when further painting advances made it possible to have more color options. This represents a classic example of how technology limitations might dictate the execution of project requirements.

Project integration

Project integration management specifies how the various parts of a project come together to make up the complete project. This knowledge area recognizes the importance of linking several aspects of a project into an integrated whole. The Henry Ford quote in the previous section emphasizes the importance of *togetherness* in any project environment. Project integration management area includes the processes and activities needed to identify, define, combine, unify, and coordinate the various processes and project activities. The traditional concepts of systems analysis are applicable to project processes. The definition of a project system and its components refers to the collection of interrelated elements organized for the purpose of achieving a common goal. The elements are organized to work synergistically together to generate a unified output that is greater than the sum of the individual outputs of the components. The harmony of project integration is evident in the characteristic symbol that this book uses to denote this area of PM knowledge.

While the knowledge areas of PM, as discussed in the preceding sections, overlap and can be implemented in alternate orders, it is still apparent that project integration management is the first step of the project effort. This is particularly based on the fact that the project charter and the project scope statement are developed under the project integration process. In order to achieve a complete and unified execution of a project, both qualitative and quantitative skills must come into play.

In addition to the standard PMBOK® inputs, tools, techniques, and outputs, the project team will add in-house items of interest to the steps presented in this section. Such in-house items are summarized as follows:

- *Inputs*: Other in-house (custom) factors of relevance and interest
- *Tools and techniques*: Other in-house (custom) tools and techniques
- *Outputs*: Other in-house outputs, reports, and data inferences of interest to the organization

The steps for a comprehensive technology transfer project execution are summarized in the subsequent sections.

Step 1: Develop project charter

Project charter formally authorizes a project. It is a document that provides authority to the project manager and it is usually issued by a project initiator or sponsor external to the project organization. The purpose of a charter is to define at a high level what the project is about, what the project will deliver, what resources are needed, what resources are available, and how the project is justified. The charter also represents an organizational commitment to dedicate the time and resources to the project. The charter should be shared with all stakeholders as a part of the communication requirement for the project. Cooperating stakeholders will not only sign-off on the project, but also make personal pledges to support the project. Projects are usually chartered by an enterprise, a government agency, a company, a program organization, or a portfolio organization in response to one or more of the following business opportunities or organizational problems:

- Market demand
- Response to regulatory development
- Customer request
- Business need
- Exploitation of technological advance
- Legal requirement
- Social need

The driving force for a project charter is the need for an organization to make a decision about which projects to authorize to respond to operational threats or opportunities. It is desired for a charter to be brief. Depending on the size and complexity of a project, the charter should not be more than two to three pages. Where additional details are warranted, the expatiating details can be provided as addenda to the basic charter document. The longer the basic charter, the less the likelihood that everyone will read and imbibe the contents. So, brevity and conciseness are desired virtues of good project charters. The charter should succinctly establish the purpose of the project, the participants, and general vision for the project.

The project charter is used as the basis for developing project plans. While it is developed at the outset of a project, a charter should always be fluid. It should be reviewed and updated throughout the life of the project. The components of the project charter are summarized as follows:

- Project overview
- Assigned project manager and authority level
- Project requirements
- Business needs
- Project purpose, justification, and goals
- Impact statement
- Constraints (time, cost, performance)

- Assumptions
- Project scope
- Financial implications
- Project approach (policies, procedures)
- Project organization
- Participating organizations and their respective roles and level of participation
- Summary milestone schedule
- Stakeholder influences
- Assumptions and constraints (organizational, environmental, external)
- Business plan and expected return on investment, if applicable
- Summary budget

The project charter does not include the project plan. Planning documents, which may include project schedule, quality plan, staff plan, communication hierarchy, financial plan, and risk plan, should be prepared and disseminated separately from the charter.

- Project overview
    - It provides a brief summary of the entire project charter. It may provide a brief history of the events that led to the project, an explanation of why the project was initiated, a description of project intent and the identity of the original project owner.
- Project goals
    - It identifies the most significant reasons for performing a project. Goals should describe improvements the project is expected to accomplish along with who will benefit from these improvements. This section should explain what various benefactors will be able to accomplish due to the project. Note that Triple C approach requires these details as a required step to securing cooperation.
- Impact statement
    - It identifies the influence the project may have on the business, operations, schedule, other projects, current technology, and existing applications. While these topics are beyond the domain of this project, each of these items should be raised for possible action.
- Constraints and assumptions
    - It identifies any deliberate or implied limitations or restrictions placed on the project along with any current or future environment the project must accommodate. These factors will influence many project decisions and strategies. The potential impact of each constraint or assumptions should be identified.

- Project scope
  - It defines the operational boundaries for the project. Specific scope components are the areas or functions to be impacted by the project and the work that will be performed. The project scope should identify both what is within the scope of the project and what is outside the scope of the project.
- Project objectives
  - It identifies expected deliverables from the project and the criteria that must be satisfied before the project is considered complete.
- Financial summary
  - It provides a recap of expected costs and benefits due to the project. These factors should be more fully defined in the cost–benefit analysis of the project. Project financials must be reforecast during the life of the effort.
- Project approach
  - It identifies the general strategy for completing the project and explains any methods or processes, particularly policies and procedures that will be used during the project.
- Project organization
  - It identifies the roles and responsibilities needed to create a meaningful and responsive structure that enables the project to be successful. Project organization must identify the people who will play each assigned role. At minimum, this section should identify who plays the roles of project owner, project manager, and core project team.
  - A project owner is required for each project.
    - This role must be filled by one or more individuals who are the fiscal trustee(s) for the project to the larger organization. This person considers the global impact of the project and deems it worthy of the required expenditure of money and time. The project owner communicates the vision for the effort and certifies the initial project charter and project plan. Should changes be required, the project owner confirms these changes and any influence on the project charter and project plan. When project decisions cannot be made at the team level, the project owner must resolve these issues. The project owner must play an active role throughout the project, especially ensuring that needed resources have been committed to the project and remain available.
  - A project manager is required for each project.
    - They are responsible for initiating, planning, executing, and controlling the total project effort. Members of the project

team report to the project manager for project assignments and are accountable to the project manager for the completion of their assigned work.

Inputs to step 1

- *Contract*: It is a contractual agreement between the organization performing the project and the organization requesting the project. It is treated as an input if the project is being done for an external customer.
- *Project SOW*: This is a narrative description of products or services to be supplied by the project. For internal projects, it is provided by the project initiator or sponsor. For external projects, it is provided by the customer as part of the bid document. For example, request for proposal, request for information, request for bid, or contract statements may contain specific work to be done. The SOW indicates the following:
  - Business need based on required training, market demand, technological advancement, legal requirement, government regulations, industry standards, or trade consensus
  - Product scope description, which documents project requirements and characteristics of the product or service that the project will deliver
  - Strategic plan, which ensures that the project supports organization's strategic goals and business tactical actions
- *Enterprise environmental factors*: These are factors that impinge upon the business environment of the organization. They include organizational structure, business culture, governmental standards, industry requirements, quality standards, trade agreements, physical infrastructure, technical assets, proprietary information, existing human resources, personnel administration, internal work authorization system, marketplace profile, competition, stakeholder requirements, stakeholder risk tolerance levels, commercial obligations, access to standardized cost estimating data, industry risk, technology variances, product life cycle, and PM information system (PMIS).
- *Organizational process assets*: These refer to the business processes used within an organization. They include standard processes, guidelines, policies, procedures, operational templates, criteria for customizing standards to specific project requirements, organization communication matrix, responsibility matrix, project closure guidelines (e.g., sunset clause), financial controls procedure, defect management procedures, change control procedures, risk control procedures, process for issuing work authorizations, processes for approving work authorizations, management of corporate knowledge base, and so on.

Tools and techniques for step 1

- *Project selection methods*: These methods are used to determine which project an organization will select for implementation. The methods can range from basic seat-of-the-pants heuristics to highly complex analytical techniques. Some examples are benefit measurement methods, comparative measure of worth analysis, scoring models, benefit contribution, capital rationing approaches, budget allocation methods, and graphical analysis tools. Analytical techniques are mathematical models that use linear programming, nonlinear programming, dynamic programming, integer programming, multiattribute optimization, and other algorithmic tools.
- *PMM*: This defines the set of PM process groups, their collateral processes, and related control functions that are combined for implementation for a particular project. The methodology may or may not follow a PM standard. It may be an adaptation of an existing project implementation template. It can also be a formal mature process or informal technique that aids in effectively developing a project charter.
- PM *information system*: This is a standardized set of automated tools available within the organization and integrated into a system for the purpose of supporting the generation of a project charter, facilitating feedback as the charter is refined, controlling changes to the project charter, or releasing the approved document.
- *Expert judgment*: This is often used to assess the inputs needed to develop the project charter. Expert judgment is available from sources such as experiential database of the organization, knowledge repository, knowledge management practices, knowledge transfer protocol, business units within the organization, consultants, stakeholders, customers, sponsors, professional organizations, technical associations, and industry groups.

Output of step 1

Project charter: As defined earlier in this chapter, project charter is a formal document that authorizes a project. It provides authority to the project manager and it is usually issued by a project initiator or sponsor external to the project organization. It empowers the project team to carry out actions needed to accomplish the end goal of the project.

Step 2: Develop preliminary project scope statement

Project scope presents a definition of what needs to be done. It specifies the characteristics and boundaries of the project and its associated products and services, as well as the methods of acceptance and scope control. Scope is developed based on information provided by the project initiator or sponsor. Scope statement includes the following:

- Project and product objectives
- Product characteristics
- Service requirements
- Product acceptance criteria
- Project constraints
- Project assumptions
- Initial project organization
- Initial defined risks
- Schedule milestones
- Initial WBS
- Order-of-magnitude cost estimate
- Project configuration management requirements
- Approval requirements

Inputs to step 2

Inputs for step 2 are the same as defined for step 1 covering project charter, SOW, environmental factors, and organizational process assets.

Tools and techniques for step 2

The tools and techniques for step 2 are the same as defined for step 1 and cover PMM, PMIS, and expert judgment.

Output of step 2

The output of step 2 is the preliminary project scope statement, which was defined and described earlier.

Step 3: Develop project management plan

A PM plan includes all actions necessary to define, integrate, and coordinate all subsidiary and complementing plans into a cohesive PM plan. It defines how the project is executed, monitored and controlled, and closed. The PM plan is updated and revised through the integrated change control process. In addition, the process of developing PM plan documents the collection of outputs of planning processes and includes the following:

- PM processes selected by the PM team
- Level of implementation of each selected process
- Descriptions of tools and techniques to be used for accomplishing those processes
- How selected processes will be used to manage the specific project

- How work will be executed to accomplish the project objectives
- How changes will be monitored and controlled
- How configuration management will be performed
- How integrity of the performance measurement baselines will be maintained and used
- The requirements and techniques for communication among stakeholders
- The selected project life cycle and, for multi-phase projects, the associated project phases
- Key management reviews for content, extent, and timing

The PM plan can be a summary or integration of relevant subsidiary, auxiliary, and ancillary project plans. All efforts that are expected to contribute to the project goal can be linked into the overall project plan, each with the appropriate level of detail. Examples of subsidiary plans are as follows:

- Project scope management plan
- Schedule management plan
- Cost management plan
- Quality management plan
- Process improvement plan
- Staffing management plan
- Communication management plan
- Risk management plan
- Procurement management plan
- Milestone list
- Resource calendar
- Cost baseline
- Quality baseline
- Risk register

Inputs to step 3

Inputs to step 3 are the same as defined previously and include preliminary project scope statement, PM processes, enterprise environmental factors, and organizational process assets.

Tools and techniques for step 3

The tools and techniques for step 3 are PMM, PMIS, and expert judgment. PMM defines a process that aids a PM team in developing and controlling changes to the project plan. PMIS at this step covers the following segments:

- Automated system, which is used by the project team to do the following:
  - Support generation of the PM plan
  - Facilitate feedback as the document is developed

    - Control changes to the PM plan
    - Release the approved document
- Configuration management system, which is a subsystem that includes subprocesses for accomplishing the following:
    - Submitting proposed changes
    - Tracking systems for reviewing and authorizing changes
    - Providing a method to validate approved changes
    - Implementing change management system
- Configuration management system, which forms a collection of formal procedures used to apply technical and administrative oversight to do the following:
    - Identify and document functional and physical characteristics of a product or component
    - Control any changes to such characteristics
    - Record and report each change and its implementation status
    - Support audit of the products or components to verify conformance to requirements
- Change control system is the segment of PM information system that provides a collection of formal procedures that define how project deliverables and documentation are controlled.

Expert judgment, the third tool for step 3, is applied to develop technical and management details to be included in the PM plan.

Output of step 3

The output of step 3 is the PM plan.

Step 4: Direct and manage project execution

Step 4 requires the project manager and project team to perform multiple actions to execute the project plan successfully. Some of the required activities for project execution are summarized as follows:

- Perform activities to accomplish project objectives
- Expend effort and spend funds
- Staff, train, and manage project team members
- Obtain quotation, bids, offers, or proposals as appropriate
- Implement planned methods and standards
- Create, control, verify, and validate project deliverables
- Manage risks and implement risk response activities
- Manage sellers
- Adapt approved changes into scope, plans, and environment
- Establish and manage external and internal communication channels

- Collect project data and report cost, schedule, technical, and quality progress and status information to facilitate forecasting
- Collect and document lessons learned and implement approved process improvement activities

The process of directing and managing project execution also requires implementation of the following:

- Approved corrective actions that will bring anticipated project performance into compliance with the plan
- Approved preventive actions to reduce the probability of potential negative consequences
- Approved defect repair requests to correct product defects during quality process

Inputs to step 4

Inputs to step 4 are summarized as follows:

- PM plan.
- Approved corrective actions are documented, authorized directions required to bring expected future project performance into conformance with the PM plan.
- Approved change requests include documented, authorized changes to expand or contract project scope. They can also modify policies, PM plans, procedures, costs, budgets, or revise schedules. Change requests are implemented by the project team.
- Approved defect repair is a documented, authorized request for product correction of defect found during the quality inspection or the audit process.
- Validated defect repair is a notification that re-inspected repaired items have either been accepted or rejected.
- Administrative closure procedure documents all the activities, interactions, and related roles and responsibilities needed in executing the administrative closure procedure for the project.

Tools and techniques for step 4

The tools and techniques for step 4 are PMM and PMIS, and they were previously defined.

Outputs of step 4

- Deliverables
- Requested changes
- Implemented change requests
- Implemented corrective actions

- Implemented preventive actions
- Implemented defect repair
- Work performance information

Step 5: Monitor and control project work

No organization can be strategic without being quantitative. It is through quantitative measures that a project can be tracked, measured, assessed, and controlled. The need for monitoring and control can be evident in the request for quantification that some project funding agencies use. Some quantifiable performance measures are schedule outcome, cost-effectiveness, response time, number of reworks, and lines of computer codes developed. Monitoring and controlling are performed to monitor project processes associated with initiating, planning, executing, and closing and are concerned with the following:

- Comparing actual performance against plan
- Assessing performance to determine whether corrective or preventive actions are required, and then recommending those actions as necessary
- Analyzing, tracking, and monitoring project risks to make sure risks are identified, status is reported, response plans are being executed
- Maintaining an accurate timely information base concerning the project's products and associated documentation
- Providing information to support status reporting, progress measurement, and forecasting
- Providing forecasts to update current cost and schedule information
- Monitoring implementation of approved changes

Inputs to step 5

Inputs to step 5 include the following:

- PM plan
- Work performance plan
- Rejected change requests
  - Change requests
  - Supporting documentation
  - Change review status showing disposition of rejected change requests

Tools and techniques for step 5

- PMM
- PMIS

- Earned value technique measures performance as project moves from initiation through closure. It provides means to forecast future performance based on past performance.
- Expert judgment

Outputs of step 5

- *Recommended corrective actions*: Documented recommendations required to bring expected future project performance into conformance with the PM plan
- *Recommended preventive actions*: Documented recommendations that reduce the probability of negative consequences associated with project risks
- *Forecasts*: Estimates or predictions of conditions and events in the project's future based on information available at the time of the forecast
- *Recommended defect repairs*: Some defects found during quality inspection and audit process recommended for correction
- Requested changes

Step 6: Integrated change control

Integrated change control is performed from project inception through completion. It is required because projects rarely run according to plan. Major components of integrated change control include the following:

- Identifying when a change needs to occur or when a change has occurred
- Amending factors that circumvent change control procedures
- Reviewing and approving requested changes
- Managing and regulating flow of approved changes
- Maintaining and approving recommended corrective and preventive actions
- Controlling and updating scope, cost, budget, schedule, and quality requirements based upon approved changes
- Documenting the complete impact of requested changes
- Validating defect repair
- Controlling project quality to standards based on quality reports

Combining configuration management system with integrated change control includes identifying, documenting, and controlling changes to the baseline. Project-wide application of the configuration management system, including change control processes, accomplishes the following three major objectives:

- Establishes evolutionary method to consistently identify and request changes to established baselines and to assess the value and effectiveness of those changes
- Provides opportunities to continuously validate and improve the project by considering the impact of each change
- Provides the mechanism for the PM team to consistently communicate all changes to the stakeholders

Integrated change control process includes some specific activities of the configuration management summarized as follows:

- *Configuration identification*: This provides the basis from which the configuration of products is defined and verified, products and documents are labeled, changes are managed, and accountability is maintained.
- *Configuration status accounting*: This involves capturing, storing, and accessing configuration information needed to manage products and product information effectively.
- *Configuration verification and auditing*: This involves confirming that performance and functional requirements defined in the configuration documentation have been satisfied.

Under integrated change control, every documented requested change must be either accepted or rejected by some authority within the PM team or an external organization representing the initiator, sponsor, or customer. Integrated change control can, possibly, be controlled by a change control board.

Inputs to step 6

The inputs to step 6 include the following items, which were all described earlier:

- PM plan
- Requested changes
- Work performance information
- Recommended preventive actions
- Deliverables

Tools and techniques for step 6

- *PMM*: This defines a process that helps a PM team in implementing integrated change control for the project.
- *PMIS*: This is an automated system used by the team as an aid for the implementation of an integrated change control process for

the project. It also facilitates feedback for the project and controls changes across the project.

- *Expert judgment*: This refers to the process whereby the project team uses stakeholders with expert judgment on the change control board to control and approve all requested changes to any aspect of the project.

Outputs of step 6

The outputs of step 6 include the following:

- Approved change requested
- Rejected change requests
- PM plan (updates)
- Project scope statement (updates)
- Approved corrective actions
- Approved preventive actions
- Approved defect repair
- Validated defect repair
- Deliverables

Step 7: Close project

At its completion, a project must be formally closed. This involves performing the project closure portion of the PM plan or closure of a phase of a multiphase project. There are two main procedures developed to establish interactions necessary to perform the closure function. They are as follows:

- *Administrative closure procedure*: This provides details of all activities, interactions, and related roles and responsibilities involved in executing the administrative closure of the project. It also covers activities needed to collect project records, analyze project success or failure, gather lessons learned, and archive project information.
- *Contract closure procedure*: This involves both product verification and administrative closure for any existing contract agreements. Contract closure procedure is an input to the close contract process.

Inputs to step 7

The inputs to step 7 are the following:

- PM plan.
- *Contract documentation*: This is an input used to perform the contract closure process and includes the contract itself as well as changes to the contract and other documentation, such as technical approach, product description, or deliverable acceptance criteria and procedures.

- Enterprise environmental factors.
- Organizational process assets.
- Work performance information.
- Deliverables, as previously described, and also as approved by the integrated change control process.

Tools and techniques of step 7

- PMM
- PMIS
- Expert judgment

Outputs of step 7

- Administrative closure procedure
  - Procedures to transfer the project products or services to production and/or operations are developed and established at this stage.
  - This stage covers a step-by-step methodology for administrative closure that addresses the following:
    - Actions and activities to define the stakeholder approval requirements for changes and all levels of deliverables.
    - Actions and activities confirm project has met all sponsor, customer, and other stakeholders' requirements.
    - Actions and activities to verify the all deliverables have been provided and accepted.
    - Actions and activities to validate completion and exit criteria for the project.
- Contract closure procedure
  - This stage provides a step-by-step methodology that addresses the terms and conditions of the contracts and any required completion or exit criteria for contract closure.
  - Actions performed at this stage formally close all contracts associated with the completed project.
- Final product, service, or result
  - Formal acceptance and handover of the final product, service, or result that the project was authorized to provide.
  - Formal statement confirming that the terms of the contract have been met.
- Organizational process assets (updates)
  - Development of the index and location of project documentation using the configuration management system.
  - Formal acceptance documentation, which formally indicates the customer or sponsor has officially accepted the deliverables.
  - Project files, which contain all documentation resulting from the project activities.

- Project closure documents, which consist of a formal documentation indicating the completion of the project and transfer of deliverables.
- Historical information, which is transferred to knowledge base of lessons learned for use by future projects.
- Traceability of process steps.

Technology sustainability

Technology project efforts must be sustained in other for a project to achieve the intended end results in the long run. Project sustainability is not often addressed in PM, but it is very essential particularly for projects involving technology transfers.

Sustainability, in ordinary usage, refers to the capacity to maintain a certain process or state indefinitely. In day-to-day parlance, the concept of sustainability is applied more specifically to living organisms and systems, particularly environmental systems. As applied to the human community, sustainability has been expressed as meeting the needs of the present without compromising the ability of future generations to meet their own needs. The term has its roots in ecology as the ability of an ecosystem to maintain ecological processes, functions, biodiversity, and productivity into the future. When applied to systems, sustainability brings out the conventional attributes of a system in terms of having the following capabilities:

- Self-regulation
- Self-adjustment
- Self-correction
- Self-recreation

To be sustainable, nature's resources must only be used at a rate at which they can be replenished naturally. Within the environmental science community, there is a strong belief that the world is progressing on an unsustainable path because the Earth's limited natural resources are being consumed more rapidly than they are being replaced by nature. Consequently, a collective human effort to keep human use of natural resources within the sustainable development aspect of the Earth's finite resource limits has become an issue of urgent importance. Unsustainable management of natural resources puts the Earth's future in jeopardy.

Sustainability has become a widespread, controversial, and complex issue that is applied in many different ways, including the following:

- Sustainability of ecological systems or biological organization (e.g., wetlands, prairies, forests)
- Sustainability of human organization (e.g., ecovillages, ecomunicipalities, sustainable cities)

- Sustainability of human activities and disciplines (e.g., sustainable agriculture, sustainable architecture, sustainable energy)
- Sustainability of projects (e.g., operations, resource allocation, cost control)

For project integration, the concept of sustainability can be applied to facilitate collaboration across project entities. The process of achieving continued improvement in operations, in a sustainable way, requires that engineers create new technologies that facilitate interdisciplinary thought exchanges. Under the project methodology of this book, sustainability means asking questions that relate to the consistency and long-term execution of the project plan. Essential questions that should be addressed include the following:

- Is the project plan supportable under current operating conditions?
- Will the estimated cost remain stable within some tolerance bounds?
- Are human resources skills able to keep up with the ever changing requirements of a complex project?
- Will the project team persevere toward the project goal, through both rough and smooth times?
- Will interest and enthusiasm for the project be sustained beyond the initial euphoria?

Figure 12.3 illustrates the author's conjecture of the span of how organizations adopt new technology. The innovators are those who are always on the cutting edge of the applications of PM. They find creative ways to use existing tools and invest in creating new and enhanced tools. The early adopters are those who capitalize on using PM whenever an

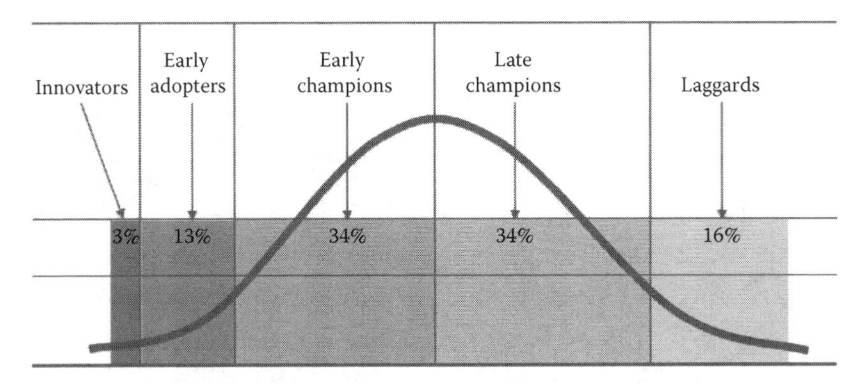

Figure 12.3 Span of technology adoption timeline.

opportunity develops. Early champions are those who provide support for and encourage the application of PM. Most managers will fall in this category. Late champions are those who say show me the money and I will believe. They eventually come around to the side of PM once they see and experience the benefits directly. The laggards are those who remain obstinate no matter what. They deprive themselves of the structured benefits of PM.

Case application for technology transfer

PM approach to technology adoption and implementation can help to resolve technology conflicts and ensure that compatible technologies are adopted for specific needs. Technology transfer and adoption should be pursued by following the scope of the life cycle of a project. This consists of conceptualization, definition, implementation, tracking and reporting, control, and phaseout given as follows:

- *Technology conceptualization*: This is the stage where a need for the proposed technology is identified, defined, and justified. It may be necessary to develop or modify a technology to fit industrialization needs.
- *Technology definition*: This is the phase at which the purpose of the technology is defined and clarified. A mission statement is the major output of this stage. The mission statement should specify what the technology transfer project is expected to achieve. For industrial development projects, a technology strategy should be developed at the national level. The mission statement for each specific project should be based on the national strategy.
- *Resource consideration*: Project goals and objectives are accomplished by applying resources to functional requirements. Resources, in the context of PM, are generally made up of people and equipment, which are typically in short supply. The labor and skill levels required to run and maintain an industrial technology should be outlined as a part of the project planning function.
- *Project implementation*: This stage involves the operational aspects of initiating the technology project in accordance with specified goals. This phase may cover a series of functions including planning, organizing, resource allocation, and scheduling. Research and development should play a crucial role in technology implementation. The early investment committed to research will later on translate into more successful technology implementation.
- *Technology tracking and reporting*: This phase entails the diagnostic process of checking whether or not the results of a technology transfer project conform to plans and specifications.

- *Technology control*: In this function, necessary actions are taken to correct unacceptable deviations from expected performance. Steps must be taken to ensure that a technology is not misused. Contingency plans should be developed for technology catastrophes in case of accidental or deliberate misuse.
- *Technology phaseout*: This is the termination stage of a technology. The termination of a technology should be executed with as much commitment as the termination of all projects associated with it. Obsolete technologies should be phased out at the appropriate times as new technologies become available. If a technology is not phased out when appropriate, the support for it will wane and it will be subject to neglect and misuse.

Technology performance evaluation

Time, cost, and performance form the basis for the operating characteristics of industrial technology. These factors help to determine the basis for technology planning and control. Technology control is the process of reducing the deviation between actual performance and expected performance. To be able to control a technology, we must be able to measure its performance. Measurements are taken regarding time (schedule), performance, and cost.

The traditional procedures for measuring progress, evaluating performance, and taking control actions are not adequate for technology management where events are more dynamic. Some of the causes of technology control problems are listed in the subsequent sections.

Technology schedule problems

- Procrastination of technology adoption
- Poor precedence relationships
- Unreliable feasibility study
- Delay of critical activities
- Hasty implementation
- Technical problems
- Poor timing

Technology performance problems

- Inappropriate application
- Poor quality of hardware
- Lack of clear objectives
- Maintenance problems
- Poor functionality

- Improper location
- Lack of training

Technology cost problems

- Lack of complete economic analysis
- Inadequate starting budget
- Effects of inflation
- Poor cost reporting
- High overhead cost
- Unreasonable scope
- High labor cost

Technology planning

Technology planning involves establishing the set of actions needed to achieve the goals of technology. It is needed to provide the following benefits:

- Minimize the effects of technology uncertainties
- Clarify technology goals and objectives
- Provide basis for evaluating the progress of technology project
- Establish measures of technology performance
- Determine required personnel responsibilities

The following list and descriptions are the set of recommended actions involved in technology planning.

- *Technology overview*: This specifies the goals and scope of the technology as well as its relevance to the overall mission of the development project. The major milestones, with a description of the significance of each, should be documented. In addition, the organization structure to be used for the project should be established.
- *Technology goal*: This consists of a detailed description of the overall goal of the proposed technology. A technology goal may be a combination of a series of objectives. Each objective should be detailed with respect to its impact on the project goal. The major actions that will be taken to ensure the achievement of the objectives should also be identified.
- *Strategic planning*: The overall long-range purposes of the technology should be defined. Its feasible useful life should be defined. A frequent problem with technology is the extension of useful life well beyond the time of obsolescence. If a technologically feasible life is defined during the planning stage, it will be less traumatic to replace the technology at the appropriate time.

- *Technology policy*: This refers to the general guideline for personnel actions and managerial decision making relating to the adoption and implementation of new technology. The project policy indicates how the project plan will be executed. The chain of command and the network of information flow are governed by the established policy for the project. A lack of policy creates a fertile ground for incoherence in technology implementation due to conflicting interpretations of the project plan.
- *Technology procedures*: These are the detailed methods of complying with established technology policies. A policy, for example, may stipulate that the approval of the project manager must be obtained for all purchases. A procedure then may specify how the approval should be obtained: oral or written. A policy without procedures creates an opportunity for misinterpretations.
- *Technology resources*: The resources (manpower and equipment) required for the adoption and implementation of new technology should be defined. Currently available technology resources should be identified along with resources yet to be acquired. The time frame of availability of each resource should also be specified. Issues such as personnel recruiting and technical training should be addressed early in the project.
- *Technology budget*: Technology cannot be acquired without adequate budget. Some of the cost aspects that will influence technology budgeting are first cost, fixed cost, operating cost, maintenance cost, direct/indirect costs, overhead cost, and salvage value.
- *Operating characteristics*: A specification of the operating characteristics of the technology should be developed. Questions about operating characteristics should include the following: What inputs will be required by the technology? What outputs are expected from the technology? What is the scope of the technology implementation? How will its performance be measured and evaluated? Is the infrastructure suitable for the technology's physical configuration? What maintenance is needed and how will the maintenance be performed? What infrastructure is required to support the technology?
- *Cost–benefit analysis*: The bottom line in any technology implementation is composed of profit, benefit, and/or performance. An analysis of the expenditure required for implementing the technology versus its benefits should be conducted to see if the technology is worthwhile. Can an existing technology satisfy the needs more economically? Even if future needs dictate the acquisition of the technology, economic decisions should still consider prevailing circumstances.
- *Technology performance measures*: Performance standards should be established for any new technology. The standards provide the yardsticks against which adoption and implementation progress may be

compared. In addition, the methods by which the performance will be analyzed should also be defined to avoid ambiguities in tracking and reporting.

- *Technology organization*: This involves organizing the technology personnel with respect to required duties, assigned responsibilities, and desired personnel interactions. The organization structure serves as the coordination model for the technology implementation project.
- *Technology work breakdown structure*: This refers to a logical breakdown of the technology implementation project into major functional clusters. This facilitates a more efficient and logical analysis of the elements and activities involved in the adoption and implementation process. A WBS shows the hierarchy of major tasks required to accomplish project objectives. It permits the implementation of the *divide and conquer* concepts. Overall technology planning and control can be improved by using WBS. A large project may be broken down into smaller sub-projects that may, in turn, be broken down into task groups.
- *Potential technology problems*: New technologies are prone to new and unknown problems. Contingency plans must be established. Preparation must be made for unexpected problems such as technical failure, software bugs, personnel problems, technological changes, equipment failures, human errors, data deficiency, and decision uncertainties.
- *Acquisition process*: Technology acquisition deals with the process of procuring and implementing a proposed technology. The acquisition may involve the acquisition of both physical and intangible assets. The acquisition process should normally cover the following analysis:
  - *Hardware*: This involves an analysis of the physical component of the technology. Questions that should be asked may concern such factors as size, weight, safety features, space requirement, and ergonomics.
  - *Software*: This relates to the analysis of the program code, user interface, and operating characteristics of any computer software needed to support the proposed technology.
  - *Site selection and installation*: A suitable and accessible location should be found for the physical component of the new technology. The surrounding infrastructure should be such that the function of the technology is facilitated.

Triple C model for technology management

The successful execution of technology projects requires a coordinated approach that should utilize conventional project planning and control techniques as well as other management strategies. Intricate organizational and human factors considerations come into play in the implementation of today's complex technologies. The success of projects depends on good levels of communication, cooperation, and coordination. The Triple C model can facilitate a systematic approach to the planning, organizing, scheduling, and control of a technology transfer project. In the context of technology adoption and implementation, the following functions should be addressed:

- Technology communication
- Technology cooperation
- Technology coordination

Technology communication

Many technologies are just emerging from research laboratories. There are still apprehensions and controversies regarding their potential impacts. Implementing new technology projects may generate concerns both within and outside an organization. A frequent concern is the loss of jobs. Sometimes, there may be uncertainties about the impacts of the proposed technology. Proper communication can help educate all the audience of the project concerning its merits. Informative communication is especially important in cases where cultural aspects may influence the success of technology transfer. The people that will be affected by the project should be informed early as to

- The need for the technology.
- The direct and indirect benefits of the technology.
- The resources that are available to support the technology.
- The nature, scope, and the expected impact of the technology.
- The expected contributions of individuals involved in the technology.
- The person, group, or organization responsible for the technology.
- The observers, beneficiaries, and proponents of the technology.
- The potential effect of the failure of the project.
- The funding source for the project.

Wide communication is a vital factor in securing support for a technology transfer project. A concerted effort should be made to inform those who should know. Moreover, the communication channel must be kept open throughout the project life cycle. In addition to in-house communication, external sources should also be consulted as appropriate. A technology consortium may be established to facilitate communication with external sources. The consortia will link various organizations with respect to specific technology products and objectives. This will facilitate exchange of both technical and managerial ideas.

A communication responsibility matrix shows the linking of sources of communication and targets of communication. Cells within the matrix indicate the subject of the desired communication. There should be at least one filled cell in each row and each column of the matrix. This assures that each individual of department has at least one communication source or target associated with him or her. With a communication responsibility matrix, a clear understanding of what needs to be communicated to whom can be developed.

Technology cooperation

Not only must people be informed, but their cooperation must also be explicitly sought. Merely saying *yeah* to a project is not enough assurance of full cooperation. In effect, the proposed technology must be sold to management and employees. A structured approach to seeking cooperation should help identify and explain the following items to the project personnel:

- The cooperative efforts are needed to assure success of the technology.
- The time frame is involved in implementing the technology.
- The criticality of cooperation to the technology.
- The organizational benefits of cooperation.
- The implication of lack of cooperation.

Workforce needs

Manpower needs for new technology implementation are usually difficult to quantify accurately due to the variability in application objectives, personnel competence, lack of precedents, and technological constraints. Requests for manpower resources should be realistically based on the practicality of the project situation. This helps to secure credibility and support for the effort. The question of personnel training should also be critically analyzed. Many technologies are still at the stage where there is a limited supply of skilled manpower. Thus, organizing a competent and cooperative project group may not be simple.

Equipment requirement

A list of physical equipment needed to support the technology project should be made. Those already available and those yet to be obtained should be identified. This will aid management in evaluating organizational capability in implementing the proposed technology. Detailed equipment documentation and explanation will contribute significantly in winning the cooperation of management and employees.

Time requirement

Due to the lack of precedents for new technologies, it is difficult to obtain accurate estimates of project schedule. When seeking the cooperation of those to be involved in the project, it will be prudent to propose conservative schedule estimates. The dynamism of technology can make time estimation to be very volatile. Setting a precarious terminal date for the project completion may provide the grounds for criticism if the deadline is not met. Time allowances should be made for technology changeover. One strategy is to present time requirements in terms of a series of milestones.

Technology coordination

Having successfully initiated the communication and cooperation functions, the efforts of the project team must, thereafter, be coordinated. Coordination facilitates conducive organization of project efforts. The development of a responsibility chart can be very helpful at this stage. A responsibility chart is a matrix consisting of columns of individual or functional departments and rows of required actions. Cells within the matrix are filled with relationship codes that indicate who is responsible for what. A responsibility chart helps to avoid overlooking critical functional requirements and responsibilities. The matrix should indicate the following:

- Who is to do what?
- Who is to inform whom of what?
- Whose approval is needed for what?
- Who is responsible for which results?
- What personnel interfaces are involved?
- What support is needed from whom for what functions?

The use of a PM approach is particularly important when technology is transferred from a developed nation (or organization) to a less developed nation (or organization). In some cases, fully completed technology products cannot be transferred due to the incompatibility of operating conditions and requirements. In some cases, the receiving organization has the means to adapt transferred technology concepts, theories, and ideas to local conditions to generate the desired products. In other cases, the receiving organization has the infrastructure to implement technology procedures and guidelines to obtain the required products at the local level.

In order to reach the overall goal of industrial technology transfer, it is essential that the most suitable technology be identified promptly, transferred under the most conducive terms, and implemented at the receiving organization in the most appropriate manner. PM offers guidelines and models that can be helpful in achieving these aims.

chapter thirteen

Nigeria's industrialization

Case example

As Nigeria is the new largest economy in Africa, it is appropriate to use case examples from Nigeria to illustrate the potentials for Africa–USA technology and industrialization interactions. While on assignment for the United Nations Development Program in the mid-1990s, the author had the privilege of working with the Nigerian Federal Ministry of Industry in Abuja, Nigeria. There, he witnessed and participated in several industrialization initiatives. His specific assignment was to develop operational guidelines for the Industrial Development Centers (IDCs) located in every state of the country. A key part of his responsibilities for the IDCs was to explore technology transfer initiatives that could benefit the country's industrial establishments. The accounts presented here are based on Sahel (1992). The accounts confirm that there is no shortage of laudable planning in developing countries. The problems are often that of sustainable implementation of the plans. This issue points to the need for new management approaches to national planning schemes. It is hoped that the contents of this book can steer policy makers, decision makers, leaders, practitioners, portfolio managers, and entrepreneurs into the right direction of better management of technology transfer engagements.

Nigeria's policy objectives of the 1980s

The overriding aim of government's development program in the Fourth National Development Plan was improvement in the living conditions of the people. A number of specific objectives were focused on this goal. The specific objectives set for the fourth plan period included the following:

- Increase in the real income of the average citizen.
- More even distribution of income among individuals and socioeconomic groups.
- Reduction in the level of unemployment.
- Increase in the supply of skilled manpower.
- Reduction in the dependence of the economy on a narrow range of activities.

- Balanced development, that is, the achievement of balance in the development of the different sectors of the economy and the various geographical areas of the country.
- Increased participation of Nigerians in the ownership and management of productive enterprises.
- Greater self-reliance—increased dependence on our domestic resources to achieve the various objectives of society. This also implied increased efforts to achieve optimum utilization of human and material resources
- Development of technology.
- Increased productivity.
- National orientation to discipline, better attitude to work, and cleaner environment.

In the context of the plan, agricultural production and processing were given the highest priority. This became necessary because of the need to feed a large and rapidly growing population without the massive importation of food and to provide the basic raw materials needed for agro-based industries. It was also a necessary strategy for the development of the rural areas so as to reduce migration from rural to urban areas. A rapid growth in agricultural productions was an essential component of the strategy of self-reliance, which was a major objective of the plan.

Education and manpower development, strengthening of economic infrastructures, and provision of health and housing needs constituted priority areas of the plans.

The fourth plan was launched at a time when the country's production of crude oil, which was the main source of government revenue and foreign exchange earnings, had virtually stabilized. As conceived in the previous third plan period, the basic strategy was that of using the resources generated by this asset to ensure expansion of the productive capacity of the economy in order to lay a solid foundation for self-sustaining growth and development in the shortest time possible. As part of the strategy for moving the economy toward greater self-reliance, the use of domestic resources, in the planning and execution of projects, was encouraged. Government agencies were similarly encouraged where appropriate, to develop in-house capacity for the execution of projects. The growth of indigenous contracting capacity was also given support especially with respect to civil engineering construction and the patronage of Nigerian companies.

Nigeria's industrialization objectives of the 1980s

Apart from the emphasis given to the establishment of basic industries and the inclusions of technology development objectives in the third and fourth national development plans, the objectives of industrial development of the 1970s continued into the early 1980s. These objectives included among others: providing greater employment opportunities, increased export of manufactured goods, dispersal of industries, improving technological skills and capability, increased local content of industrial output, attracting foreign capital, and so on.

The major goal of government in the industrial sector in the 1980s was to encourage and promote directly and indirectly rapid development of manufacturing and allied activities. In order to ensure that industrialization brought in its wake a truly beneficial economic and social development, industrial development was guided by the following objectives:

- *Manufacturing value added* (MVA): Industries with low local inputs into manufacturing contribute minimally to national economic development. In order to encourage meaningful economic development, industrial enterprises that utilize local inputs were encouraged. Particular attention was given to
  - Utilization of local raw materials
  - Agro-based industries
  - Industries with linkage effects
  - Industries with backward integration potentials
  - Manpower development
  - Technology development
- *Industrial research and development (R&D)*: The continued improvement of the quality of industrial products, the production process, and the amount of local input are desirable attributes that industries should pursue through R&D. In addition to the research institutions established by government, private sector industrial establishments were encouraged to maintain effective R&D units that can find solutions to the problems facing industries.
- *Employment*: The industrial sector was expected to generate remunerative employment for large numbers of people throughout the country. Within the limits of economic viability, industries were encouraged to adopt labor-intensive technologies. Industries based on materials available in each locality or making the most of local skill and manpower enjoyed more governmental incentives.
- *Export promotion*: Apart from satisfying domestic demand, Nigerian industries were also expected to produce goods and services for export. To engage meaningfully in international trade meant that Nigerian industrial products should be internationally competitive in terms of both quality and price.
- *Industrial dispersal*: The dispersal of industries throughout the country was aimed at reducing the concentration of industries in few areas. It was also designed to check rural–urban migration and promote balanced physical development.

With rapid deterioration in the balance of payment since 1982, tariffs were increased, quantitative controls were extended to a wide range of imports, and import licensing and the allocation of foreign exchange were introduced. This highly protective regime insulated domestic industries from foreign competition and encouraged the growth of import-based consumer goods and assembly industries with little domestic value added.

Structural adjustment programme and industrialization

The introduction of structural adjustment programme (SAP) was the culmination of various attempts to contain the structural defects that became apparent in the economy as a result of the collapse of oil prices in 1981. The fall in oil prices led to a fall in Nigeria's foreign reserve from about N 5.8 billion by mid-1981 to under N 1.2 billion by early 1982. Real Gross Domestic Product (GDP) fell by 1.9% in 1982 and 6.3% in 1983. Attempts to secure a $2.0 billion standby facility from the International Monetary Fund (IMF) and to raise loans through syndication by a consortium of banks failed because *a certificate of good economic health and management* was required. To achieve the requirements of the IMF as far back as 1982, government introduced the 1982 Stabilization Act, which brought a number of austerity measures including some of the IMF requirements, that is, reduction in budget and balance of payments deficits, reduction in government spending, and cutting subventions to parastatals. However, the government did not agree to a devaluation of the naira.

The Economic Recovery Programme announced in the 1986 annual budget had the following major objectives:

- To restructure and diversify the productive base of the economy in order to reduce dependence on the oil sector and imports
- To achieve fiscal and balance of payments viability over the medium term
- To lay the basis for sustainable noninflationary growth over the medium and long term.

In order to achieve the above objectives, the following policy elements were built into the program:

- Strengthening existing demand management policies
- Adoption of a realistic exchange rate policy
- Further rationalization/restructuring of the customs tariffs to aid the promotion of industrial diversification
- Simplification of the regulations and guidelines governing industrial investment and commercial banking activities
- Adoption of appropriate pricing policies especially for petroleum products and public utilities

The 1986 budget, in effect, gave the basic outline of SAP that was introduced later in the year. The core policies introduced involved three main approaches, given as follows:

- Actions to correct the over valuation of the naira (introduction of the second-tier foreign exchange market [SFEM])
- Actions to overcome the observed public sector inefficiencies (rationalization/privatization and deregulation)
- Actions to relieve the debt burden and attract a net inflow of foreign capital while keeping a lid on foreign loans

Exchange rate

Prior to the advent of SAP, the exchange rate of the naira vis-à-vis the major convertible currencies was fixed by the Central Bank of Nigeria (CBN) allowing only for a few minor adjustments. Its allocation among contending demands was by import licensing. Apart from its arbitrariness and administrative rigidity, the import licensing procedure was fraught with corruption, maladministration and led to the misallocation of scarce resources.

The introduction of the SFEM brought in a measure of market forces determination of the exchange rate. Between September 1986 and April 1987, there were two foreign exchange markets, the first-tier that is a fixed rate, to which a crawling peg was applied in order to slide it down until the gap between it and the second-tier (SFEM) was closed.

Three modifications were introduced into the operation of the market:

- The first was the merging of the first and second tiers in April 1987.
- The second was the introduction of the *Dutch Auction* system in April 1987 whereby the banks pay whatever they bid, instead of paying the emerging marginal rate for the day.
- The third shift came in January 1989 which enforced the requirements that autonomous funds be sold at the rate that emerges from the auction.

Trade policies

Before SAP, import procedures were a maze of regulations. Along with SFEM came the abolition of import licensing. The 30% import duty surcharge introduced under the emergency decree was also abolished. Custom duties were streamlined and made more uniform. This culminated in the Customs and Excise Tariff Consolidation Decree of 1988. This decree gave a 7-year life for specified rates in order to give certainty to business planning. The prohibition list was equally reduced to 16 items.

The devalued naira has enhanced the naira value of traditional nonoil exports. The SFEM decree allowed nonoil export revenue to be retained 100% by the exporter in his domiciliary account and to convert proceeds in such account into naira through the interbank foreign exchange market.

Fiscal policies

It was expected that in line with the entire adjustment process, government expenditure would be trimmed and budgetary deficits held down to within 3% of GDP. To achieve this, there was to be

- Restraints on growth of public wage bills.
- Emphasis on maintenance culture rather than on starting new projects.
- Privatization or commercialization of certain parastatals and government-owned companies.
- Allocation of capital expenditure to key projects that are near completion and that have greater impact on other sectors.
- Emphasis on market-determined prices: subsidies were to be removed, especially those on petroleum and energy.

However, these have not been easy to apply. In 1987, the deficit went up to almost 10% of GDP despite increased naira revenue from oil and apparent increase in domestic taxation. Efforts have, however, been made to implement a number of these measures.

The introduction of SAP in 1986 was inevitable given the palpably dismal situation in the first half of the 1980s. However, the efficacy of the program given the excruciating demand that it has made on the individual and on social structures need to be assessed. In the implementation of the reforms, various targets are set, though they were reviewed periodically.

In 1987 and 1988, the credit ceilings for both the private and public sectors were exceeded by very wide margins. In the succeeding year, however, actual performance was far below the target for both the private and public sectors. Though actual performance in 1990 was higher than the target for both sectors, the divergence in relative terms was much smaller than for the earlier years. In the same vein, the targets for money supply were exceeded by wide margins especially in 1988 and 1990. This situation persisted in spite of the several mopping up operations of the CBN. The inability to keep to the money supply target or the targets for credit does not reflect the tight monetary posture of the central bank.

The major objective in the deregulating interest rate determination is to ensure efficient allocation of resources. Implicit in this assumption is the expectation that real interest rates would be positive. In this respect, performance has been partially successful as interest rates for 1988 and

1989 were decisively negative. However, the nominal interest rates rose significantly during the period.

The determination of the exchange rate by market forces was to remove or minimize price distortions, thereby effecting a more efficient allocation of resources. The exchange rate has depreciated the naira value significantly since the introduction of SAP. The act appears to have been overplayed since the naira is now generally believed to be grossly undervalued. International price comparison, which the massive depreciation was to correct, has thus become ridiculous in terms of the worth of naira.

Though the investment climate has improved tremendously with the liberalization of procedure for the repatriation of profit and new industrial policy, both domestic and foreign investment on new projects have been marginal. Replacement costs let alone new investment have become virtually impossible given the new interest and exchange rate structures. Restoration of confidence of foreign investors in the Nigerian economy does not appear to have been achieved.

Corporate planning has also become increasingly difficult because of the relative volatility of both the interest and the exchange rate.

A target budget deficit not exceeding 3.5% of GDP was set under SAP. However, the figure has not been achieved since the inception of the program. This is in spite of massive reductions in subsidies for social services and for consumer products such as petroleum products as well as fertilizer.

Foreign debts have increased from between $22 billion and $25 billion in 1985 to about $33 billion in 1990 reflecting the effect of rescheduling and further drawdown on existing loans.

Reliance on market forces is the major thrust of SAP. The deregulation of the interest rate, the introduction of the foreign exchange market, and the reduction of the effective rate of protection in the industrial sector are in line with this thrust. However, the market itself imposes a constraint on the development process. Credit and foreign exchange are more readily available to large-scale enterprises and borrowers because short-term economic activities are favored over activities with long gestation period. In short, the market is bedeviled with greater imperfections than before and this impedes growth and development.

Nigeria's policy objectives of the 1990s

The government had introduced in 1989 a new industrial policy document that will chart a new course for the industrial sector in the 1990s. The document brought together various measures adopted under SAP, which have an impact on industrial development. It noted that a major problem of the industrial sector was inadequate supply of imported inputs and spare parts, resulting in gross underutilization of installed capacity.

Other problems plaguing the sector, according to the document, include geographical concentration, high production costs, low value added, high import content, and low level of foreign investment.

The policy focus will therefore be to

- Encourage the accelerated development and use of local raw materials and intermediate inputs rather than the dependence on imports.
- Develop and utilize technology.
- Maximize the growth in value added of manufacturing production.
- Promote export-oriented industries.
- Generate employment through the encouragement of private sector small- and medium-scale industries.
- Remove bottlenecks and constraints that hamper industrial development.
- Liberalize controls to facilitate greater indigenous and foreign investment.

To achieve these objectives, government would continue to put in place

- A realistic and flexible exchange rate policy that will reflect the scarce nature of foreign exchange and therefore ensure its efficient allocation.
- Tariff structures that will ensure effective protection of industries.
- Incentives that would improve the investment climate.

The role of the public sector has been streamlined as follows:

- Encouraging increased private sector by privatizing government holdings in existing industrial enterprises
- Playing a catalytic role in establishing new core industries
- Providing and improving infrastructural facilities
- Improving the regulatory environment
- Improving the investment climate prevailing in the country
- Establishing a set of industrial priorities
- Harmonizing industrial policies at federal, state, and local government levels

Expected roles of the private sector

The failure of past industrial development policies has, since the mid-1980s, led to the search for alternative strategies. Federal Ministry of Information (FMI), therefore, in 1988 adopted a new blueprint for industrial development, which gave prominence to the role of the private sector. The FMI subsequently sought the assistance of UNIDO to put in place a new management approach to industrial development under an industrial master plan (IMP) otherwise referred to as strategic management of industrial development (SMID). In this new approach, the industrial landscape is seen as a system, describing a network of relationships, among various actors. The industrial system, using criteria such as resource base, production process, and market orientation, is subdivided into subsystems for operational purposes. The objective is to identify key subsystems whose development would provide a catalytic role in the development of the whole system.

In giving effect to this new management approach, the FMI in August 1988 established the National Committee of Industrial Development to collaborate with UNIDO in executing this new management approach.

Strategic management of industrial development

The IMP or SMID is predicated on the need to organize a network of actors around an industrial activity with a view to having a comprehensive and perspective view of the investment problems in that particular line of industrial activity. This network of actors is referred to as the Strategic Consultative Group (SCG). For a given industrial activity around which an SCG is organized, membership is drawn from manufacturers, raw material suppliers, transporters, policy makers, providers of infrastructural support services as well as distributors of the industrial goods/services.

Each SCG is expected to examine the problems of investment in its subsystem against the background of the existing policy environment, technology, structures of production, and market potentials. The SCGs are then expected to evolve a workable investment program toward restructuring, re-orienting, or developing new product lines, which would ensure greater efficiency and competitiveness of the subsystem in both the domestic and international markets.

It should be noted that in the past, a major weakness of previous industrial and indeed overall development plans was that projects were planned, sectoral policies were articulated, and incentives were put in place without adequate consideration of the effect of one project on another, or the disincentive effects of an incentive policy for activities that are not covered by the particular incentive structure. In addition, policies and programs—due to political instability and/or political expediency—were subject to rapid changes. Worse still, many of the policies and programs of government were put in place under a near total absence of current facts and figures on the Nigerian economy, and without sustained interaction and consultation with target beneficiaries of policies.

The macroeconomic implications of the above situation are rather too obvious to warrant much elaboration. We only need to note the offsetting tendencies of inconsistent policies and incentives on the economy as a whole, as well as the mutual suspicion between public and private sector operators.

The SMID otherwise known as the IMP seeks to minimize the problems of policy and program inconsistencies in the development of a nation's industry. This, the IMP seeks to achieve through a systematic policy articulation that clearly spells out the inter- and intrasectoral relationships within industry and between industry and other sectors of the economy including, of course, the external sector. It also seeks to modernize and rationalize existing industries in order to enhance their efficiency and competitiveness, as well as set up the institutional framework for the strategic management of the industrial sector. Strategic management facilitates a clearer grasp of the intersectoral resource flows and their implications. It encourages and promotes constant interaction and consultation between various actors in the economic scene, especially between public sector policy makers and private sector operators. The interactive and consultative processes lead to mutual perception of problems, objectives, and strategies and thus minimize conflicts that usually arise in the legitimate pursuit of various group interests. In a way, the IMP strategy seeks to temper the mood of society with objective realities of market forces and vice versa. It greatly facilitates information flow thus helping to reduce transactions costs.

In short, IMP can be described as a framework that provides for a dynamic and a regulated flow of investment funds and therefore a regulated industrial development in an environment of deregulation. The regulation is exercised through a system of industrial incentives and institutional support which then allow domestic and foreign entrepreneurs to interact, to invest, and to operate in areas that they calculate will maximize their returns. A careful and sustained monitoring of the responses of the private entrepreneurs and the associated resources flows to policy signals, which in turn enables government to adjust the incentive structure and institutional framework in an effective manner. In a way, therefore, the IMP is an attempt to promote medium- to long-term investment through the provision of clear, articulate, and *negotiable* industrial policy. The negotiation is undertaken through organized consultations between the private and public sectors in the course of the plan formulation and its implementation. The key objectives of the industrial master plan can be summarized as follows:

- Achievement of an orderly and coordinated industrial development
- Provision of a stable environment and a negotiated incentive structure necessary for medium- to long-term investment by the private sector
- Domestication of the industrial process through the promotion of local sourcing of industrial raw materials and the development of domestic technological capability

- Enhancement of economic efficiency so as to improve Nigeria's international competitiveness in the export of manufactured goods and thereby increase the country's export earnings and reduce the burden of debt
- Full development and exploitation of Nigeria's potentially large domestic market
- Provision of a flexible industrial base capable of quick but non-disruptive adjustments to national and international shocks
- Diversification and restructuring of the industrial base—diversification strengthens the economy by providing a cushion of shocks, that is, slacks in some economic activities are offset by positive developments in other areas, while restructuring seeks, in particular, to correct the imbalance between investment in consumer and capital goods production
- Maximization of the benefits of industrial development through well-designed intersectoral linkages, the benefits being, in particular, employment, outputs, material welfare, and some degree of national prestige

Over the short-term, therefore, the IMP seeks to correct the shortcomings of the structure of industry and identify problems that must be resolved over the medium- and long-term periods so that the long-term objectives are attainable. Industrial priorities and the sequence of industrial development are also specified over this period. For example, the over dependence on imported raw material and poor infrastructures are short-term problems; effective domestication of industrial structure will be pursued, for example, domestication of technology and industrial skills and the realization of the full benefits of industrial development, guided by the vision of industrial development already specified in the short-term.

Strategies for implementation

In outlining the modalities for the preparation and implementation of the IMP, it should be noted that the focus on the private sector as the moving force in industrial development and the implied decrease in the role of government in direct productive investment increase rather than diminish the complex role of the government in the management of the economy. It will require enhanced public sector professional and technical skills as well as developed institutional and political frameworks.

The implementation of the IMP involves three main phases. The first phase involves the formulation of the strategic guidelines for the IMP and the setting up of the institutional framework for the strategic management of the industrial sector. Phase two involves the formulation of strategic investment and action programs for industrial development, while phase three is to mobilize resources to implement and to monitor the investment and action programs.

In other words, the three interrelated phases imply first identifying and analyzing the problems of the industrial sector. Based on the analysis of the problems identified, design the appropriate investment strategies with given levels of investment related to time-specific growth targets. In order to ensure that investments are undertaken and targets are met, resources need to be mobilized and a strategic management and monitoring of the investment program put in place. The constant monitoring of the responses of private sector investments to given incentives provides the guide to appropriate adjustments when necessary. The unique advantages to these three interrelated phases of the exercise can be summarized as follows:

- The prior study of the chosen areas of focus and the consensus between all actors regarding an enhanced position of knowledge and facts of the economy and a mutual understanding of each other's roles.
- The formulation of investment plans based on an earlier problem identification and analysis makes for a clearer articulation of financing gaps. This in turn provides an objective basis to both investors and financiers to reach a common ground on terms and conditions of lending.
- The institutional framework for monitoring the investment programs, namely, the SCG, ensures timely and mutually agreed modifications to policies and programs in a non-disruptive manner.

It is hoped that the new SMID would become a working reality that will put industrial development in Nigeria on a firmer footing through active participation and indeed leadership of the private sector.

Problems of implementation

From the planning experience discussed above, four major drawbacks in Nigeria's planning systems are apparent. These are as follows:

- Framework of objectives, priorities, and strategy is not often properly grounded in social reality. The planners do not show commitment to or identification with the vast majority of Nigerians: this in turn means that there is little or no attention paid to the plan.
- The macroeconomic framework used is unduly simplistic. It is often based on anticipated money flows, and the aggregate plan is derived from the basic model in which public sector leads and private sector follows. There is no direct link between projects and those that will implement them.

- Project selection process is so poor and disjointed that the exercise often appears as merely filling in the plan, and not what is expected to be implemented on the ground.
- The link between plan formulation and plan implementation leads to a situation where the plan merely lists the objectives at which policy measures will be directed without providing details on the implied relationships between policy instruments and the various target variables.

Specifically, the plans showed that some of the various structural weaknesses of Nigeria's industrial sector are

- Heavy import dependence because of the low stage of development of intermediate goods and capital goods industries.
- Inadequate linkages among the various industrial subsectors and between industry and other sectors such as agriculture, mining, and construction.
- Technological dependence because of the rudimentary engineering industries.

Nigeria's strategies of the 1960s

During the colonial era, Nigeria did not have any explicit policy on industrialization. The colonial administration regarded the colonial territory as a source of supply of raw materials and one to serve as market for industries in Britain. It only embarked on minimum processing of natural resources to avoid carrying unnecessary financial burden. This was why the Ten-Year Plan of Development and Welfare (1946–1955) did not envisage any real industrialization program.

By 1950, there were only 20–30 factories that could pass for industries. These included palm oil milling, palm kernel crushing, groundnut crushing, cotton ginning, leather tanning, sawmilling, beer brewing, and oil seed milling among others. All these accounted for no more than 2.7% of the GDP. Therefore, one can safely say that Nigeria's industrialization started with primary processing of local raw materials before export.

Between 1955 and 1966, more local resource-based industries to supply the local market were developed. These included soap making, soft drinks, tobacco/cigarettes, bakeries, and confectioneries. This approach followed the classic theory of industrialization that advocates that newly industrializing countries should start by studying their imports and try to substitute the products by local manufacturing, beginning from simple consumer goods to more complex consumer durables. This gave birth to the import substitution or resource-based strategy, which was adopted in the first national development plan (1962–1968). Many of the industries were small-scale industries as revealed by the industrial survey of 1963, which reported only 160 establishments employing 100 persons or more.

The import substitution strategy favored consumer goods industries that relied heavily on imported machinery and raw materials. The imported items were brought in with minimal or no duty, while the industries themselves were protected by high tariffs as *infant industries.*

During the first decade of independence, the Nigerian government did not have much capital, and therefore, investment in industry was left largely to foreign-dominated private sector. However, a number of industrial ventures were started under the sponsorship of some federal agencies and the regional development boards. Some of these industries include cement plants, breweries, cocoa processing, and steel rolling mills. In addition, market forces, fiscal incentives, and intensified program of import substitution motivated the establishment of medium- and large-scale manufacturing plants besides agricultural processing industries. The manufacturing sector growth, though still small in relation to GDP, had an annual growth rate of 15%–20% between 1960 and 1965.

The import substitution strategy has the following structural weaknesses:

- Over reliance on imported inputs
- A bias for the production of consumer goods
- Predominately internal market oriented for its output
- Over-dependence on protection from external competition which bred inefficiency and market distortions
- Capital intensive method of production
- Over-concentration on secondary stage processing with little or no internal linkages in the economy

The increasing pressures brought about by the implementation of the first national development plan, as well as the increasing social and political burdens of the newly independent administration, led to a call for foreign investment. Unfortunately, the response of foreign investors to this attempt to assert economic independence was poor and nonchalant.

In spite of Nigeria's many natural attractions to investors such as extensive market, natural resources, and enormous amount of entrepreneurial talent, factors had tended to discourage investment in the manufacturing sector. The most important of these factors included uncertainty about the policy environment, a restrictive and cumbersome regulatory framework, and inadequate incentives.

Nigeria's strategies of the 1970s

In the 1970s, the growth of manufacturing was fairly rapid, averaging about 13% during the period. Consumer goods accounted for over 70% of total MVA, while the balance of less than 30% was accounted for by the intermediate and capital goods. The decade saw the emergence of the public sector as the leader in industrial investment and development, hence a singular emphasis on public sector planning with little or no attention paid to the private sector.

The public sector investments in the capital goods industry arose out of the need to shift the emphasis of the economy from outward to inward looking, that is, internal reliance for bother capital goods and overall economic development.

Some of the disadvantages of the internal market-oriented import-substitution approach are as follows:

- It encourages production of consumer goods with limited linkage effects within the economy.
- It perpetuates, through foreign private industrial investment and the gains to investors, the national disadvantage of international division of labor.
- There is negligible advancement in industrial technology.
- It promotes foreign domination.
- An unorganized industrial development (a stagnating and small industrial sector).
- The high protective tariffs that usually support the strategy have had unfavorable effect on the structure of the manufacturing industry as it tends to weaken the incentive to make good quality products and to improve productivity.
- It gives weak linkage effects that are responsible for the drift from the rural to the urban areas.

The implications of some of the above disadvantages include the following:

- Lack of capacity to produce capital goods
- Low level of acquired technical skill
- Perpetual dependence on external sources for the solution of Nigeria's basic problems

Some of the advantages of the strategy of import substitution are as follows:

- Impressive growth in gross output, value added, and employment
- Dominant source of growth, accounting for about 80% of the growth in the output of manufacturing industries

The import substitution strategy of industrialization was continued during the second national development plan period. However, because of large oil revenues received in the early 1970s, the country added a strategy of public sector-led industrialization. These two planks in our industrialization effort had their advantages and disadvantages, although later developments in the economy tilted the balance in favor of the advantages, thus showing up the structural weaknesses of the nation's industrial sector.

However, at the beginning of the 1970s, the manufacturing sector was still being dominated by foreigners. This development, which was considered unsatisfactory, led to the formulation of indigenization policy as enunciated in the Nigerian Enterprises Promotion Decree (NEPD) of 1972.

Nigerian Enterprises Promotion Board

The NEPD was promulgated in 1972 (amended 1977) to involve Nigerians in the ownership, control, and management of certain enterprises. The degree was designed to

- Create opportunities for Nigerian indigenous businessmen.
- Raise the proportion of indigenous ownership of industrial investments.
- Maximize local retention of profits.
- Raise the level of local production of intermediate goods.
- Increase Nigerian participation in decision making in the management of larger commercial and industrial establishments.
- Advance and promote enterprises in which citizens of Nigeria shall participate fully and play a dominate role.
- Advise the minister of industries on clearly defined policy guidelines for the promotion of Nigerian enterprises.
- Advise the minister on measures that would assist in ensuring the assumption of the control of the Nigerian economy by Nigerians in the shortest possible time.
- Determine any matter relating to business enterprises in Nigeria generally in respect of commerce and industry that may be referred to it in accordance with any direction of minister, and to make such recommendations as many as necessary on those matters in such manners as may be directed by the minister.

All enterprises were classified under three schedules, each of which indicates the maximum equity participation a foreigner could hold in the enterprises listed under it. The list under each schedule has been modified from time to time to reflect developmental priorities.

During the period of its existence, the Nigerian Enterprises Promotion Board played a critical role as a promoter of indigenous enterprise. However, one of the major weaknesses of the decree was that the foreigners quickly realigned their enterprises to seek for exemption. Therefore, from the outset, the board faced the problems of

- Identifying the enterprises affected by the decree.
- Recruiting adequate number of qualified staff to man the various organs of the system.
- Finding necessary finance to acquire the enterprises affected.
- Training the required manpower to replace the alien owner–managers in Schedule I enterprises.

As a result of the above shortcomings, the 1972 decree was reviewed and the 1977 NEPD was promulgated to correct them. The features of the new decree designed to increase its effectiveness were as follows:

- Every enterprise was affected.
- State promotion committees were created to supplement the activities of the board in their respective states.
- Zonal offices were opened to increase operational effectiveness of the Inspectorate Department.
- The Nigeria Bank for Commerce and Industry (NBCI) and Nigeria Industrial Development Bank (NIDB) as well as other banks and financial institutions were specifically mandated to give loans to prospective Nigerian investors.

During the 1980s, it was observed that total investment as a share of GDP had fallen due to several factors ranging from inadequate foreign capital inflow to low levels of internal savings. This situation led government to review the investment environment particularly in light of the fact that Nigerian entrepreneurs have come of age and are able to hold their ground in various types of enterprises. The review was also necessitated by the fact that demands of SAP imposed the spirit of competition and efficiency in production and quality of goods and prices acceptable to consumers.

Therefore, in order to encourage foreign capital inflow, government amended the NEPD of 1977. With the amendment, there now exists only one list of scheduled enterprises exclusively reserved for Nigerians for the purpose of 100% equity ownership. All other businesses not contained in the schedule are now open for 100% Nigerian for foreign participation except in the areas of banking, insurance, petroleum prospecting, and mining, where the previous arrangements still subsist. This new ownership structure applies to new investments only as well as financing and incentives, trade information, trade facilities, export publicity, and training.

Despite the existence of the Nigerian Export Promotion Council (NEPC), the following structural problems militated against export promotion:

- Poor product quality
- Uncompetitive prices
- Inadequate transportation
- Supply bottlenecks

Therefore, in order to further encourage the export of agricultural and manufactured products, the federal government promulgated the Export Promotion Decree (No. 18) of 1986. The decree provided for the following incentives:

- Retention of export proceeds in foreign currency in a *domiciliary* bank account in Nigeria. Such money may be used to pay for specified activities.
- Abolition of export license requirements for the exportation of manufactured or processed goods.
- Export Credit Guarantee and Insurance Scheme.
- Scrapping of commodities boards which by their monopolistic nature hindered the operations of free market forces.
- Export Development and Expansion Funds to assist exporters cushion some of their initial expenses and to boost their future consignments.
- Export Adjustment Scheme Fund which is a form of subsidy on cost of production.
- Re-discount of short-term bills for export.
- Additional capital allowance where applicable.
- Tax relief on interest on capital invested in export-oriented industries.
- Export-free zones in selected states of the federation.

Industrial training fund

The Industrial Training Fund (ITF), established by Decree No. 47 of 1971, is the body responsible for promoting and encouraging the acquisition of skills in industry and commerce. It is expected to continue to generate indigenous trained manpower to meet the needs of the economy. To this end, the fund provided facilitates for training of persons employed in industry and commerce, approves courses, and appraises facilities provided by other bodies. It also assists individual persons or corporate organizations in finding facilities for training for employment in industry and commerce and conducts or assists others to conduct research into any matter relating to training in industry.

Strategies of the 1980s

At a meeting held in 1980, the African heads of state adopted the Lagos Plan of Action designed to promote the twin goals of promoting self-reliance in food production and self-sustaining industrialization. In order to facilitate the industrial component, the heads of state declared in the 1980s the Industrial Development Decade for Africa (IDDA). The IDDA programs operationalized the concept of self-sustaining industrialization to include

- Indentifying and establishing core industries such as iron and steel, petrochemical fertilizer, and pulp and paper projects.
- Reassessing industrial strategies toward a local resource-based industrialization.
- Creating internal engines of growth.

The objectives of the Lagos Plan of Action and IDDA are therefore to

- Reduce dependence on external demand stimuli.
- Reduce dependence on external factor input or other supply.
- Internalize employment and income multiplier effects of investment.

We have noted earlier that the industrialization policies of the 1960s and 1970s have been mainly geared toward the promotion of import substitution and the manufacture of consumer goods. Although import substitution is not fundamentally bad, it should not have been predicated upon the importation of raw materials and components and should not, as is often the case, be a mere assembly operation that contributes neither to the upgrading of indigenous resources nor to the development of technological potentials. The strategy also prolonged the dependence on external sources while the creation of capital intensive import-substitution industries distorted cost structures. For emphasis, it is necessary to again restate those other recurring structural weaknesses of the industrial sector that include the following:

- Lack of base from which Nigeria could launch into a self-sustaining industrial growth since there was virtually no engineering industries or well-developed capital goods sector, and was therefore vulnerable to external shocks.
- Few or no real linkage industries, or intersectional transactions.
- Lack of requisite stick of human capital with the necessary technological skills for an industrial takeoff.

Policy package

In order to revitalize the manufacturing sector and to accelerate its pace of growth and development, government adopted an industrial policy package, which was intended to

- Increase industrial value added.
- Diversify the industrial base by supplying basic and intermediate inputs.
- Generate productive employment and raise productivity.
- Increase the technological capacity.
- Increase private sector participation in the economy.
- Disperse industries evenly across the country.

The objectives of the fourth national development plan were therefore designed to tackle the issues mentioned above. However, the plan was more or less stillborn as the economy was then beset with problems that had their origin largely in the worldwide recession, and partly in poor economic management. By 1985, the economy was virtually on the verge of collapse.

Capacity utilization of most industries was below 20% owing to lack of foreign exchange for raw materials and spare parts, and inflation had attained an intolerable level. It therefore became clear that a structural reform was inevitable. It was in this context that the SAP was introduced in July 1986.

Nigeria's industrial policy of 1980s

A revised National Industrial Policy was launched in 1980 with the following objectives:

- Maximization of local value added
- Promotion of research and development
- Employment generation
- Promotion of export-oriented industries
- Industrial dispersal

The strategies for achieving the objectives were as follows:

- Full recognition of private enterprise and initiative as the responsibility of the state for the welfare of every citizen
- Provision of adequate incentives to industries

Then, industrial policy could not be fully implemented as a result of the oil crisis in 1981, which led to a dwindling of Nigeria's foreign reserve.

When the military took over in 1984 following the economic crisis of 1982–83, the administration looked inward and took some decisive steps to discipline the populace by curbing their extravagant lifestyles that depended heavily on imports. The government refused to reopen negotiations with IMF and preferred austerity and prudent economic management to fight the problem.

When in Babangida administration assumed office in August 1985, it introduced a period of economic emergency that was to last for 15 months. The decree granted the present wide powers to deal with the problems of the economy. The government also sought and used to introduce an economic recovery program that aimed at altering and realigning aggregate domestic expenditure and production patterns so as to minimize dependence on imports, enhance nonoil export base, as well as bring back the economy on the path of stead and balanced growth.

The 1986 budget gave the main outline for the SAP that was published later that year.

The problems of industry were specifically analyzed and provided for in the SAP document. It noted that given the structure of Nigeria's industries, its major problem was inadequate supply of imported inputs and spare parts, resulting in gross underutilization of installed capacity.

Government therefore came to the conclusion that local sourcing would be the long-term solution, since foreign exchange will never be adequate. The SAP document provided specifically for the industrial sector as follows:

- Encouraging the accelerated development and use of local raw materials and intermediate inputs rather than depend on imported ones
- Development and utilization of local technology
- Maximizing the growth in value added of manufacturing production
- Promoting export-oriented industries
- Generating employment through the encouragement of private sector small- and medium-scale industries
- Removing bottlenecks and constraints that hamper industrial development including infrastructural, manpower, and administrative deficiencies
- Liberalizing financial controls to facilitate greater indigenous and foreign investment

Structural adjustment programme

The major objectives of the SAP are as follows:

- To restructure and diversify the productive base of the economy in order to reduce dependence on the oil sector and on imports
- To achieve fiscal and balance of payments viability over the medium term
- To lay the basis for a sustainable, noninflationary growth over the medium and long term

For the industrial sector, the strategy under the program aimed at

- Encouraging the accelerated development and use of local raw materials and intermediate inputs rather than depend on imported ones.
- Development and utilization of local technology.
- Maximizing the growth in value added of manufacturing production.
- Promoting export-oriented industries.
- Generating employment through the encouragement of private sector small- and medium-scale industries.
- Removing bottlenecks and constraints that hamper industrial development including infrastructural, manpower, and administrative deficiencies.
- Liberalizing controls to facilitate greater indigenous and foreign investment.

In order to achieve these objectives, the government put in place a number of measures that have materially changed the macroeconomic framework in which industries have to operate. Major aspects of the new environment include the following:

- Emphasis on private sector-led growth
- A measure of deregulation of erstwhile controls on economic activities
- Liberalization of trade but providing minimum protection to strategic industries
- Freer access to foreign exchange market where the exchange rate for naira is determined by the interplay of market forces
- Privatization and commercialization of some government investments
- A more open, competitive economy
- Abolition of commodity boards
- Abolition of the Import Licensing Scheme
- Removal or substantial reduction in various government subsidies by means of commercialization of some utility services such as electricity supply and telecommunications services

Impact of SAP on manufacturing activities

It is necessary to recapitulate the economic climate under which SAP was introduced in July 1986, a gloomy background of mounting external debt, unhealthy investment, and the failure of the regime of stringent trade and exchange controls, which had been pursued in the previous two decades.

As a consequence of SAP, many large enterprises in agro-industry diversified into the growing and processing of primary agricultural produce for further processing into raw materials for their plants.

Also, due to improved access to foreign exchange and consequently to imported inputs, capacity utilization of the industrial sector increased on the average from 30% in 1986 to 37% in 1987 and 40% in 1988. Very high capacity utilization was experienced in domestic resource intensive activities such as textiles, furniture, rubber, and nonmetallic mineral products, while low capacity utilization was found among import-oriented activities with relatively low domestic value added such as car assembly, electrical and electronic plants, basic chemicals, and pharmaceuticals.

In conclusion, the decade of the 1980s witness a deep crisis in the nation's industrial development. The crisis is evidenced by

- A steady decline in capacity utilization.
- Deterioration of the tools of production and decay of capital assets in some cases.
- Low rate of investment in industry and even in some cases, disinvestment.
- High cost, low quality, and hence uncompetitive products.

In the 1980s, Nigeria's industrial landscape consisted of industrial ventures that were ridiculously dependent on imports, with crushing effect on balance of payments as well as the stock of scarce foreign exchange. These and more were the trend SAP sought to reverse.

Nigeria's strategies of the 1990s

There was a deep crisis in our industrial development during the 1980s. As a result, the aims and objectives of the *first* IDDA were largely unfulfilled, and they have been re-stated for the *second* decade of the 1990s. Some of the lessons learned during the 1980s included the following:

- The need to reappraise policies and strategies
- The need to reduce dependence on external factors for our development, and as a corollary
- The need to create internal engines of growth so as to internalize the multiplier effect of investment

Therefore, the experience of the 1980s and of past development planning efforts convinced the government of the need for a change in the development planning strategy for the country. A plan for the industrial sector would define the sector's forward and backward linkages with other sectors of the economy such as agriculture, transportation, construction, communication, mining, and energy. This led government to decide on the mechanism of a perspective plan and three-year rolling plans, so as to put in place *plans that are subject to periodic reviews rather than fixed plans.* The plan is expected to forge a closer link with the annual budget.

Perspective plan

Nigeria has adopted a 15–20-year perspective plan for the purpose of taking a long-term structural view of the economy. The plan was launched in 1990 with the following objectives:

- To specify long-term socioeconomic development goals and targets
- To articulate the developmental options open to the nation over the longer term
- To identify possible bottlenecks and problems in the way of achieving these developmental goals and targets
- To articulate policies and strategies for eliminating the bottlenecks and effectively pursuing the long-term goals and targets
- To specify sectoral investment priorities, and importantly
- To ensure consistency in the management and trajectory of the economy over the long term

The key objectives of the plan are as follows:

- Attainment of higher levels of self-sufficiency in the production of food and other raw materials
- Laying of solid foundation for a self-reliant industrial development as a key of self-sustaining dynamic and noninflationary growth, and promoting industrial peace and harmony
- Creating ample employment opportunities as a means of containing unemployment problems
- Enhancing the level of sociopolitical awareness of the people and further strengthening the base for a market-oriented economy, and mitigating the adverse impact of the economic downturn and the adjustment process on the most affected groups

Rolling plans

The strategy of the rolling plan is to consolidate the achievements that have been made so far from the SAP. A major aspect of the SAP is the creation of the appropriate policy environment that would promote the growth of the direct production sectors of the economy, that is, agricultural and industrial sectors through effective mobilization of available development resources, promotion of in-flow of foreign investment and efficiency in resource allocation through a pricing system that responds appropriately to market signals.

The first of the three-year rolling plans, 1990–1992, echoed the Lagos Plan of Action and the SAP objectives. It stated inter alia that the objectives of self-sufficiency in the production of food and agro-based raw materials is in line with the current efforts to expand the productive base of the economy.

The rolling plan 1990–1992 has identified the following major constraints on the industrial sector:

- Shortage of industrial raw materials and other inputs
- Infrastructural constraints
- Inadequate linkage among industrial subsectors
- Administrative and institutional constraints

Whereas the rolling plan and the perspective plan give indications of qualitative and quantitative changes envisaged for the economy, it was still considered necessary to have sectoral master plans that will contain details of how those changes are to be programmed and implemented. As a result, government decided to embark on an industrial master plan.

Nigeria's industrial master plan

As stated in the First National Rolling Plan (1990–1992), the industrial master plan is "aimed at promoting the development of an efficient industrial system through the determination and definition of all the functional aspects of an industrial system, and the preparation of an action plan to achieve established objectives and targets" (Iwuagwu 2011).

In this case, an IMP is merely a tool for the SMID. It provides the basis for guiding investment into productive facilities, support services, and training.

The developmental issues to be addressed by the IMP would include

- Local resources as factors input.
- Technological issues.
- Manpower development for industrialization.
- Physical and institutional infrastructure for industrial development.

The strategy of industrial master plan has now been translated into what is called SMID. Like the master plan, the SMID is a tool for management. SMID sets in motion the various processes, institutions, and stages leading to the strategic choices of a path for the industrial development. It calls for re-orientation on the proper role of the state in industrial development, especially in the context of a market-oriented, open competitive system implied by SAP.

In adopting the new strategy, the government has started to withdraw from certain areas of productive investment, through its privatization and commercialization program.

Also, through various policies, the government is installing a more open and de-regulated market economy.

Move toward privatization

The strategy of the privatization/commercialization of public enterprises was adopted in July 1989 by the federal government.

Privatization as a public policy was toyed with in the early 1980s. The Onosode Commission set up by the Shagari Administration in 1982 and a study group set up by the Buhari Administration in 1984 recommended the privatization of certain parastatals. However, the issue did not evoke much public debate until President Babangida in his 1986 Budget Speech announced government's decision to adopt a privatization policy. Subsequently, in July 1988, the federal government promulgated the Privatization and Commercialization Decree No. 25 of 1988, which provided the legal framework for the implementation of the program within the context of the ongoing restructuring of the Nigerian economy. The decree empowers the Technical Committee on Privatization and Commercialization to implement the program.

The basic objectives of the program are to

- Restructure and rationalize the public sector in order to lessen the dominance of unproductive investment in the sector.
- Re-orientate the public enterprises for privatization and commercialization toward a new horizon of performance improvement, viability, and overall efficiency.
- Ensure positive returns on public sector investment in commercialized enterprises.
- Check the present absolute dependence on the treasury for funding by otherwise commercially oriented parastatals and so encourage their approach to the Nigerian capital market.
- Initiate the process of gradual handing over to the private sector such public enterprises which, by their nature and type of operation, are best performed by the private sector.

The enterprises affected are spelt out in Schedule I (privatization) and Schedule II (commercialization) of the decree. The listings are in the following four categories:

- *Category I*: Thirty-seven enterprises in which equity held by the federal government shall be partially privatized.

- *Category II*: Sixty-seven enterprises in which 100% of the equity held by the federal government shall be fully privatized.
- *Category III*: Fourteen enterprises billed for partial commercialization.
- *Category IV*: Eleven enterprises to be fully commercialized.

In all, a total of 129 enterprises are affected by the exercise. Some of the apprehensions of the policy include the fact that

- The exercise could be compromised by interest groups and political considerations thereby replacing public monopoly with private monopolies.
- Workers in the privatized companies are likely to suffer certain disadvantages, because most of them are unlikely to be able to buy equity shares, their remuneration could be reduced, and their job security could be jeopardized.
- It could open the floodgate for foreign investment, thus eroding the gains of the indigenization exercise of the 1970s.

Other issues that have been raised on the program relate to

- Control, especially of enterprises in which government relinquishes its entire shareholding (i.e., 100% privatization) without the presence of a core group of shareholders with adequate knowledge or experience of the particular business.
- Ensuring adequate compensation for government investment by way of equity and loans having regard to the age of such public sector investments.

Nigeria industrial policy of 1989

The new industrial policy was put in place in 1989. The policy document brings together various measures that had been adopted under SAP, which has an impact on industrial development. It accepts the fact that the realization of the objectives of accelerated industrial development hinges on the response of the private sector to the new set of policies.

The role of the public sector has been streamlined as follows:

- Encouraging private sector participation by privatizing government holdings in industry
- Playing a catalytic role in establishing new core industries
- Improving the regulatory environment
- Improving the investment climate prevailing in the country

- Establishing a clear set of industrial priorities
- Harmonizing industrial policies at federal, state, and local levels of government

To achieve the above objectives, government is reforming the institutional framework that regulates industrial investment in the country. Some of these institutions are discussed subsequently.

Industrial development coordinating committee

One of the problems that is used to militate against foreign investment in Nigeria was the restrictive and cumbersome regulatory environment under which the economy operated. As a result, government in formulating its new industrial policy established an institutional arrangement for investment promotion through Decree No. 36 of 1988 which established the IDCC. It is an interministerial committee to ensure speedy decision making in matters concerning new investments in industry.

The committee was set up to improve the investment climate and attract investors by centralizing and simplifying procedures for issuing all pre-investment approvals. It is a one-stop agency for all pre-investment approvals. This approach removes duplication, delays, and frustration that tended to scare away genuine investors.

Small-scale industries

The present industrialization strategy is predicated on the development of local resource-based industries and small-scale industries.

- A close look at the structure of the manufacturing sector shows that a large percentage of the industries produce consumer goods that are heavily dependent on imported basic and intermediate raw materials. For example, at the 1987 Foreign Exchange Market (FEM) bidding, about 40% of the foreign exchange disbursement went to the importation of raw materials for manufacturing industries, and about 35% to the importation of machinery and spare parts including completely knocked down parts. As a result, government has decided to encourage industrial activities that rely heavily on resources that are abundantly available locally. This strategy will increase the multiplier effect of the industrialization process. Government will also encourage local production of intermediate goods and the fabrication of capital goods (of intermediate technology content) as well as spare parts.
- In the past few years, attention has been focused on the merits of small-scale industries. These include employment generation, greater utilization of local financial and material resources, promotion of

indigenous technology, tremendous opportunities for subcontracting by the large-scale industries, and forward and backward linkages. Also, the initial capital required for investment is low; and the capital/labor and capital/output ratios are relatively low. Therefore, a rapid industrial development must be based on the development of small-scale industries. In the pursuance of this strategy, the local government areas in the country have been grouped into three zones based on criteria such as per capita industrial output, the degree/extent of social and economic infrastructures available, and the level of development of the labor market. The three zones are as follows:

- *Zone 1*: Industrially and economically developed local government areas
- *Zone 2*: Less industrially and economically developed local government areas
- *Zone 3*: Lest industrially and economically developed local government areas

Government also intends to make all areas of the country attractive to new investors through a package of incentives, including a program of industrial layouts and craft villages' development. The government will also assist state governments with matching grants in the establishment of industrial estates for small-scale industries. Other activities include the ongoing Entrepreneurial Development Programme, the Working-For-Yourself Programmes, and the Training the Trainers Scheme. These programs show great promise of developing the corps of entrepreneurs needed for successful implementation of the small-scale industrialization strategy.

Industrial incentives

These industrial development strategies are complemented by a number of incentives, which have been put in place by the government to encourage both indigenous and foreign investors. The existing incentives can be classified as follows: fiscal, trade/exchange rate policies, and export promotion measures. Others include special assistance programs, the provision of infrastructural facilities, and extension services at industrial estates as well as manpower training. Some of these are as follows:

1. *Fiscal incentives*: Fiscal incentives are meant to provide for deductions and allowances in the determination of taxes payable by manufacturing enterprises. The existing fiscal incentives are as follows:
   a. *Pioneer status*: The Income Relief Act of 1958 amended by Decree 22 of 1971 provides that public companies be granted specific tax holiday on corporate income. The objective of this incentive is to

encourage industrialists to establish industries that the government considers critical to the overall development of the country. It is also intended to attract foreign investment to Nigeria for the purpose of promoting industrial expansion and the development of the country's natural resource. The scheme discriminates in favor of industries with bias for

i. Export-oriented activities.
ii. Labor intensive processes.
iii. Local raw material sourcing.
iv. The development of infrastructural facilities.
v. On-the-job training.

The relief covers a nonrenewable period of 5 years for pioneer industries, and 7 years for those of them located in economically disadvantaged areas.

b. *Tax relief for research and development*: Industrial establishments are expected to undertake R&D activities for the improvement of their processes and products. To this end, up to 120% of expenses on R&D are tax deductable so long as such research activities are carried out within the country, and are related to the activities of the company. There is a higher allowance of 140% for the development of local raw materials. In addition, the total capital expenditure might be written off against profits where the R&D is on a long-term basis.
c. *Companies income tax*: The Companies Income Tax Act 1979 with its amendments is meant to encourage potential and existing investors by reducing the corporate tax rate to 40% from 1987. It also modifies the capital allowance rates to reflect changes in rates.
d. *Tax-free dividends*: Under the specified conditions laid down by government, an individual, or company deriving dividends from any company effective from 1987 shall enjoy tax-free dividends for a period of 3 years. The tax-free period shall be 5 years if such companies are engaged in agricultural production, the processing of agricultural products, the production of petrochemicals, or the production of liquefied natural gas.
e. *Tax relief for investments in economically disadvantaged local government areas*: Investors in economically disadvantaged local government areas are entitled to the following tax incentives:
   i. Seven years income tax concessions under the Pioneer Status Scheme
   ii. Special concessions by relevant state governments
   iii. Additional 5% over and above the initial capital depreciation allowance under the Company Income Tax Act (Accelerated Capital Depreciation)

2. *Trade/exchange rate policies*: Government has put in place a new tariff regime that provides for a considerable degree of protection consistent with the industrial development objectives of developing economies at a similar state of development.

Institutional framework

In an effort to restructure the economy, there exists certain institutional mechanisms that have direct consequences for the catalytic role that government must play under the new dispensation. These institutions include the following:

1. *Industrial Inspectorate Department (IID)*: IID was established by Decree No. 53 of 1970. The functions of the IID are as follows:
   a. Certifying the actual values of capital investments in buildings, machinery, and equipment of various companies
   b. Certifying the value of imported industrial machinery for the purpose of granting approved status to nonresident capital investment
   c. Monitoring of the Comprehensive Import Supervision Scheme to ensure that the operations are in the spirit of the relevant agreement
2. *Standards Organization of Nigeria (SON)*: The SON is charged with surveillance over the products of Nigerian industries and over imported products to ensure that they meet national and international standards. SON was established to arrest the poor quality of local products, which has been traced to the raw materials and the machinery/equipment as well as the absence of quality planning and control procedures.
3. *Policy Analysis Department (PAD)*: PAD undertakes policy analysis necessary for the evaluation and effectiveness of industrial policies. These include the following:
   a. Assessing the extent of anti-export bias that is related to the trade and exchange rate regime
   b. Evaluating Nigeria's export incentives schemes
   c. Evaluating other prevailing export regulations and their administration
   d. Assessing the impact of measures in terms of government fiscal revenues, volume of import and export, and in light of industrial growth in the short and medium terms
   d. Developing fiscal investment incentives
4. *Industrial Development Coordinating Committee (IDCC)*: Introduced in June 1986 as a result of under delays and other problems, which created disincentives for many genuine investors, the IDCC now serves

as a central agency where all the approvals required for foreign investment in Nigeria can be obtained. It replaced the multiplicity of approving bodies. The objectives of establishing the IDCC are as follows:

a. To obviate delays in granting approvals for the establishment of new industries by creating one approval center that will replace the multiplicity of approving centers, which, in the past, had been responsible for unnecessary and avoidable costs to prospective investors before approvals are granted
b. To advise on policy review proposals as they relate to industrial development
c. To ensure adequate coordination and objectivity in the nation's industrial development efforts
d. To make recommendations on pertinent industrial policies including tariffs and various incentive measures aimed at enhancing steady industrial development
e. To ensure that industrial location is consistent with government's environmental policies

5. *Investment Information and Promotion Centre*: This center provides information on procedural matters and industrial incentives. It also advises and guides prospective local and foreign investors on most aspects of their investment proposals.
6. *Industrial Data Bank*: This bank is responsible for storing and retrieving data in order to provide information on existing industries in the various subsectors. Such information includes production capacities, capacity utilization and expansion plans, production costs, the extent of the market, price movements, and raw material availability in various parts of the country.
7. *National Office for Technology Acquisition and Promotion (NOTAP)*: NOTAP, initially known as National Office of Industrial Property (NOIP), was established by Decree No. 70 of 1979 to facilitate the acquisition of technology in Nigeria. The major objective of the office is to ensure that the acquisition of technology brings about social and economic gains in the areas of rapid industrialization, the development of local technologies, and the exportation of same. It is also to monitor all registered agreements to ensure that the implementation process complies with laid down terms as stipulated in the clauses of the contract. The information generated from the monitoring exercise is utilized by government in the formulation of new policies and modification of existing ones.

    In order to achieve government objectives in the technological development of the country, the office provides information and advisory services to Nigerian entrepreneurs to enhance the development of their negotiations, skills, and expertise in the acquisition of technology. It has also directed the activities of the industrial sector toward adaptive research, effective training, and the maintenance culture in all aspects of technology.

One can say that NOTAP has been able to provide effective support and assistance to Nigerian enterprises in their efforts to acquire technology at reasonable terms and conditions for rapid industrialization. Also, the activities of NOTAP have led to further improvement in the negotiating skills of entrepreneurs, while interactions with entrepreneurs have proved advantageous in leading to a reduction in serious defects in technology agreements.

Further recognizing the role of NOTAP in Nigeria's technology advancement pursuits, Chapter 14 is a contribution from the Director of NOTAP, addressing Nigeria's science, engineering, technology, and innovation system for sustainable development.

Reference

Iwuagwu, Obi, *The Cluster Concept: Will Nigeria's New Industrial Development Strategy Jumpstart The Country's Industrial Takeoff?*, *Afro-Asian Journal of Social Sciences*, Vol. 2, No. 4, pp. 1–24, 2011.

Sahel Publishing, *Industrialization in Nigeria: A Handbook*, Nigeria Federal Ministry of Industry and Technology, Sahel Publishing & Printing Co. Ltd, Lagos, Nigeria, 1992.

chapter fourteen

Technology system for sustainable development*

Umar B. Bindir

Introduction

We can define sustainable development (SD) as a mode of human development in which resource use aims to meet human needs while preserving the environment so that these needs can be met not only in the present, but also for generations to come.

Though science, engineering, technology, and innovation (SETI) is the main sectoral subject matter of this lecture (Bindir 2012), it cannot practically be singled out in isolation to lead the nation to attain the desired SD without successes in all the other economic sectors of the economy. But it is very important to emphasize at this introductory stage that SETI is the key dynamic factor driving improvements in all sectors of any economy for SD.

Leveraging science, engineering technology, and innovation

This chapter on Leveraging on the Science, Engineering Technology and Innovation (SETI) System in Nigeria for Sustainable Development will follow a format where the aspect of *leveraging* will be discussed as presenting the challenges that we must overcome, the aspect of the *SETI system* will present the appropriate deployment of available resources to provide the solutions, while Nigeria will be the subject matter on which this chapter will focus. These are all targeted at attaining a *sustainable development of the economy*. As such the central emphasis of the lecture to connect all these will be based on the transformation agenda of the present administration, which simply aims at propelling Nigeria to become one of the top 20 economies of the world as very well captured in the Vision 20-2020 document of the federal government.

* Adapted from a public lecture and contributed by Umar B. Bindir, *Leveraging on the Nigerian Science, Engineering Technology and Innovation (SETI) System for Sustainable Development*, Nigeria's Academy of Engineers (NAE), Lagos, Nigeria, April 2013.

In general terms, the objectives of mounting a transformation on any economy would target the following:

1. The total eradication of absolute poverty in the country so that citizens would unfailingly have access to all basic necessities of life
2. Ensuring that citizens have access, not just to basic needs of life, but also to all the opportunities to live a happy, secured, and fulfilling life, by every man, woman, and child in the country
3. Development levels where the country would emerge as a knowledge and a learning society built on values of hard work, honesty, discipline, productivity, sincerity, unity, and a collective sense of purpose

These are not new in civilization, as records/literature (Hargroves and Smith 2005) have made us to learn that most countries still maintain these efforts to achieve the targets, which justifies why countries establish formidable knowledge acquisition systems (universities, polytechnics, research institutions, knowledge-based industries, etc.) to acquire and apply engineering and technology for SD. The efforts and the successes globally are what today justify the classification of countries into developed (first world), emerging (second world), developing, and under-developed (third world). Nigeria still and has always been a third world country based on this classification.

At this point, it is pertinent to ponder on one definition of *the third world* by the Organisation for Economic Co-operation and Development (OECD 1998, p.12) who *daringly* defined it as follows:

> World 3 (the third world) refers to those countries—Bangladesh and Nigeria are examples—that are in severe straits with no clear path to a positive future

Can you believe it! Actually, I believe that all people of Bangladesh and Nigeria *might* not accept this *derogatory* definition. Indeed, it must be *very* annoying to all.

Nigerians, I am sure, irrespective of our orientation or local/indigenous alignment, will spontaneously wonder why this should be attributed to this great country. In actual sense, any patriotic Nigerian would want to have the opportunity to sentimentally *settle scores* with the authors of such material. But, equally, the statement should also (in my opinion) make us reflect, ponder, and deeply think of *Why This*? Especially coming from OECD, an organization composed of mainly the first world countries, many of whom are *our friends*!

Well, in terms of SD, a more sober reflection on the definition mentioned above would reveal that Nigeria, since independence in 1960 (53 years ago), despite its endowment with a large population, and abundant human and natural resources (including large deposits of oil and gas), has *disappointingly* failed to give the world any of the following among many others:

1. Globally branded products from its indigenous knowledge and industrial efforts (such as Nokia products of Finland, Apple products of the United States, Proton vehicles of Malaysia, Marcopolo buses from Brazil, Tata products from India, LG products from South Korea, and Finacle Software from India)
2. Multinational companies with outputs based on locally acquired knowledge and technology (such as Nestle from Switzerland, Friesland Campina from the Netherlands, Procter & Gamble from the United States, Infosys from India, and Indorama from Indonesia)
3. Global technical and managerial expertise bred from its knowledge system and applicable in international industry
4. Globally respected intellectual property rights (IPRs) licensed to industries internationally and earning foreign currency and technical recognition

It may further be of interest to note the numerous Nigeria's *inability* to take any local or international development policy (Indiginisation Policy, Structural Adjustment Program (SAP), Green Revolution, Operation Feed the Nation, Poverty Reduction Strategy Process (PRSP), Vision 2010, Millennium Development Goals (MDGs), National Economic Empowerment and Development Strategy (NEEDS), New Partnership for Africa's Development (NEPAD), etc.) to any clear logical and measurable conclusion to adequately develop its economy.

These could be pointers to the way Nigeria is perceived internationally; especially regarding the way we progress competitively (as a developing nation) to deliver SD using quality knowledge, engineering, and technology. This, to me, is a *big* problem that points to the need for more serious policy thoughts, strategy, and program of actions to be mounted. But as Albert Einstein said, "We can't solve problems by using the same kind of thinking used when we created them."

This indicates that there is the need for serious and drastic changes and possibly a total transformation in the way we do things, if we are to achieve tangible and measurable progress in transforming our economy. This (I believe) is the main justification for the present administration to mount the transformation agenda and targeting the attainment of the Vision 20-2020.

This lecture will unwind some of the mysteries of these failures/challenges and suggest some *viable* solutions for a positive transformation in our country, as we tackle many sectoral issues using our SETI system for SD of the economy.

Nigeria's realities

It is very clear to me that we need a strategy. But experts in strategic thinking and management would *always* admit that an early and careful review of any system using the traditional elements of strength, weakness, opportunities, and threats (SWOT) analysis would help in revealing not only the key challenges, but also actions to be taken and possible consequences useful for improvements or any positive change in the system. Relevant to this lecture, a brief SWOT analysis would reveal the following about Nigeria:

1. A population of nearly 200 million people is both a strength and an opportunity.
2. The *abundantly* available natural resources including oil and gas are both strengths and opportunities.
3. A poverty incidence that ranges between 60% and 70% (accounting for over 100 million people) is a clear weakness and a threat.
4. The dominance of *weak* rural life among the people is both a weakness and a threat.
5. The prevailing poor state of physical infrastructure is a weakness.
6. The weak socioeconomic system prevailing is actually a threat.
7. The specifically weak education system is both a clear weakness and a threat.
8. The low performance of *local* industries is a weakness.
9. Challenges of high incidence of corruption and indiscipline commonly attributed to *bad leadership* are both weaknesses and threats.
10. The maturing democracy within our present polity certainly presents a strength and also an opportunity, and so on!

Despite the "gloomy" SWOT review above, Nigeria naturally still desires to attain SD with economic attributes that should include the following:

1. A modernized economy envied by all
2. A well-focused productive country that will limit consumption of foreign things
3. A credible global visibility where we are highly respected
4. A global competitive industry that will ensure that we too take advantage of other economies
5. A dynamic population where skills and competence leads to instant jobs after graduation
6. A viable knowledge economy that will be dynamically innovative

To attain the *game changes* as listed above, it is important to understand that, all of them heavily depend on the strength, energy, and vibrancy of the National Science, Engineering, Technology and Innovation System. This is the master key anywhere and at anytime for any sustainable national development.

The national science, engineering, technology, and innovation system

There are several ingredients or recipe needed to evolve a functional and viable SETI system in any country. The main elements are given as follows:

1. A sound education system from the basic up to the tertiary levels
2. An excellent scientific research and development (R&D)
3. A dynamic intellectual property generation and management system
4. A functional and effective technology transfer mechanism
5. A strong base in traditional technologies, techniques, and skills
6. A competitive and dynamic environment for innovation

To further comprehend the dynamics of SETI, a closer and incisive look into the four elements of the system (SETI) is necessary. The key characteristics of the four elements are provided in the subsequent sections.

Science

If *science* were to be a house or an office, the dwellers and workers here are mainly scientists. In most cases, they are very highly trained and well qualified. It takes a reasonably sustained high-level training and time to breed these dwellers/workers because they have PhDs and many grow further to become academic professors. They are high wage oriented and target technical details on the *hows*, *whys*, *and wheres*. Their tools of trade are in laboratories, test stations, and workshops, which are unique, delicate, and expensive; in most cases, they have very clear concepts and detailed knowledge of the subject matter and their outputs in most cases are high-level knowledge published in various forms and at various levels of quality as a result of which they further produce more high-level knowledge workers with PhDs. These outputs are their joy, because their work (where relevant) are even internationally acclaimed, leading to promotion to higher academic positions and ranking in the form of becoming well-respected subject matter professors, with their names and work cited in other technical documents (demonstrating global mastery of the subject), which further leads to the desire of other scientists to partner, cooperate, and network with them. Due to its high specialization, an estimated 5%

of the people in most countries are involved in this phase. However, these (on their own) do not weigh/demonstrate much *relevance* when gauged on the scale of economic development of any nation. These scientific outputs must be taken to the next level in a knowledge value chain to be more effective, useful, and relevant. This leads us to the *house* of engineering and technologies.

Engineering and technology

Engineering and technology are conveniently lumped together here to ease the discussions. In this *house*, the personnel are mainly engineers, technologists, and many other relevant technocrats. They are also very well trained but have the attribute of using specialized tools of trade that are of high value, in addition to being highly skilled, to transform the high value knowledge (produced in the house of science) into proven concepts, solutions, and indications of products, processes, and skilled operational know-how. The outputs here are generally referred to (in concrete industrial terms) as intellectual properties (IPs). At this level, concepts proven will normally be in the form of prototypes, pilots, and confident field and technology deployment skills. The outputs (IPs) again generally are reclassified into Industrial Property and Copyright protectable IP. This is the application language that must be understood if such outputs are to be identified and applied appropriately. In this phase, participation of an estimated 15% of the people in most committed countries is involved. Again, the relevance of these outputs to dynamic and real-time economic development is limited if not further developed. Further value addition is needed for the investments so far to be *useful*, which is what leads us to the *house* of innovation.

Innovation

This is the *house* where value is added to IPs (engineering, technology, skills, and know-how) to evolve products and processes that are viable, competitive, and dynamic. In this phase, skillful business planning, marketing, and product development skills dominate the activities. The IPs would transform to assets protected by specific IPRs, skills to transfer the IPs (technology transfer) to industry dominate this phase. Anybody with business skills and entrepreneurial drive can participate in this category. This is a phase where an estimate of 80% of people in most countries can actively participate. The innovation phase is the most critical phase in any dynamic and real-time chain of sustainable economic development system. This is where jobs, wealth, respect, recognition, and so on become more visible and effective.

SETI dynamics summary

The review/analysis above is brief, condensed, and *highly* simplified, but even at that, it is evident that for the economy of any country to develop, it is very strategic to understand that investments in science to evolve technologies are what enable an economy to have a strong and virile knowledge system (knowledge economy). Subsequent investment on the knowledge will propel the technologies to drive innovation generating jobs and creating wealth, which is commonly referred to as knowledge-based economy. Countries that have successfully decoded, developed, and applied these have demonstrated to others that have not that the master key behind the success in a developed economy is SETI.

For countries where the above arguments are well understood and used, the science, engineering, and technology elements are well connected and networked, which propels innovation and prosperity that ultimately lead to SD. Where the science, engineering, and technology are not synergized, innovation is either weak or nonexistent resulting in low economic development, a situation that is commonly found in developing countries, and Nigeria is not an exception.

Transformation agenda, Vision 20-2020, and sustainable national development

As clearly stated in the Nigeria Vision 20:2020 document/Economic Transformation Blueprint (The National Planning Commission 2009), "the Vision 20:2020 is an articulation of the long-term intent to launch Nigeria onto a path of sustained social and economic progress and accelerate the emergence of a truly prosperous and united Nigeria. Recognizing the enormous human and natural endowments of the nation, the blueprint is an expression of Nigeria's intent to improve the living standards of her citizens and place the country among the Top 20 economies in the world with a minimum GDP of US$900 billion and a per capita income of no less than US$4,000 per annum." The ideals of this agenda is first a national task based on commitment and patriotism by all, in the sense that we know the club of the top 20 economies, and we have the capability to imagine what is needed to create an economy with a gross domestic product (GDP) of US$900 billion by the year 2020. Achieving this literally points to a collective resolution by all Nigerians (leaders and followers) that there can be no more business as usual. In part, it means that Nigeria will take advantage of the merits of knowledge, especially of engineering and technology to drive the economy so that we can achieve the development status to be among the top 20 economies in the world. A brief reflection on this desire and the current top 20 economies in the world reveals the following:

1. That the top 20 economies currently are well known and are found in North America, Europe, South America, Asia, and Oceania.
2. In each of those countries, their National SETI Policies are well defined, focused, practical, and implementable and aligned to moving *all sectors* of the economy forward.
3. Their SETI system is vibrant.
4. They have successfully bred well synergized and networked institutional knowledge systems to innovate (NSI).
5. Each country sustains a respectable R&D investment.
6. Where there are natural raw material endowments, downstream chains of value addition are highly localized and vibrant.
7. Each of those countries implements prioritized technology acquisition, with clear SETI mentor to drive it.
8. Each of those countries maintains a strong IP system to create, protect, and deploy.
9. Intense deployment of ICT characterizes the automation mode in every sector in those countries.
10. The countries all demonstrate viable Research—Industry Linkage models including the establishment of Intellectual Property and Technology Transfer Offices, speedy evolution of hi-tech companies, efficient technology incubation, establishing numerous science and technology parks, research parks, Innopolis, and Technopolis.

In all the above, the commitments and intensity of a functional SETI system is quite clear.

SETI analysis in Nigeria

In the pursuit of driving the economy to attain the Vision 20-2020 through the transformation agenda, it is worth taking a cursory look at the Nigerian SETI system so that an objective analysis can be carried out on some of the required actions. The following are visible:

1. That the knowledge infrastructure is elaborate, currently composed of 128 universities (National University Commission 2013); 138 polytechnics, monotechnics, and colleges of agriculture, 281 colleges of health technology, VEIs, and technical colleges; 63 colleges of education over 500 institutions composed of research

institutes, innovation agencies and policy implementation departments, many world class companies, large pool of knowledge workers including capacities in the diaspora.

2. A presence at every level of science, engineering, and technology bandwidth including high levels such as space science, intermediate levels including numerous machinery and processes, low levels including unit operation-based creativities, and at traditional level that is wide and numerous.
3. That Nigeria actually has all that is needed to adequately transform and become the largest economy in Africa.
4. But despite (1), (2), and (3) above, the economy is still technologically very weak with a national poverty incidence averaging 70% implying that over 100 million Nigerians are dwelling under the poverty line in absolute terms.
5. The prevailing situation in (4) above depicts that Nigerian knowledge system finds it difficult to improve people's quality of lives indicating that a *SETI valley of death* exists.
6. Many reasons might be responsible for the existence of this valley of death, but certainly include the following reasons:
    i. A *weak* technology-based industrialization strategy (policy, planning, etc.)
    ii. Lacking innovation support (venture capital/technology risk/angel funds, etc.)
    iii. A weak policy focus on relevant technology acquisition and transfer
    iv. A weak IP creation and management system
    v. A weak technology-based entrepreneurship drive (product and process development, management expertise)
    vi. A weak strategic business planning skills (feasibility, business plans, branding)
7. Nigeria maintains a high appetite for consuming imported products and locally produced items based on foreign technology on which the low capacity to absorb and domesticate such technologies is very clear.
8. The summary of all the above is that the Nigeria science system does not seem to be *aggressive* enough in moving to support the acquisition and application of the needed engineering and technology to transform the economy through innovation, resulting in high consumption of foreign technology-based innovations. Literally, the SETI system in Nigeria demonstrates all known symptoms of functional weakness including high incidence of poverty, weak physical and social infrastructure, and very low production capacity especially among the indigenous industries. These are the key indicators for the desire to transform the economy.

Political support for SETI

Since 2007, the political leadership in the country has sustained their campaign for radically transforming the Nigeria economy and improving the quality of lives of Nigerians. From the time of the late president Umaru Musa Yar'Adua when the Vision 2020 was first launched based on the implementation of a 7-point agenda to the current time of President Goodluck Jonathan where the same vision is maintained to be attained through the implementation of a transformation agenda, the objective has basically remained toward evolving an improved Nigerian economy where

1. Absolute poverty among Nigerians is totally eradicated.
2. Every man, woman, and child in the nation would have access, not just to basic minimum needs, but to all the opportunities to lead a happy and fulfilling life.
3. Nigeria strives to emerge as knowledge oriented and learning society built on values of hard work, honesty, transparency, sincerity, discipline, and a collective sense of purpose.

The leadership also sustained all indicators for its desire to ensure that SETI is entrenched in the process of transforming the economy. These are great signals and are clear pointers toward the evolution of a functional SD system, in which the traditional *triple helix* concept will become a reality in Nigeria:

1. That the government as an apex helix will provide the political, economic, moral, and technological leaderships.
2. The energy in (a) above will facilitate for the knowledge helix to mature to provide the needed knowledge and technology.
3. The Industry helix will naturally respond to both (1) and (2) above to ensure that the economy becomes productive, sustaining the generation of gainful and high wage jobs, and that will ensure sustained wealth creation in the economy.

These are all possible to attain, especially as evidences abound both locally and internationally.

The way forward

In order for the desired transformation to manifest riding on the current efforts of transformation agenda of government, it must be based on the change strategy of re-oriented and well-documented action plans that are result based have clearly set targets, performance measurement, and close monitoring and evaluation. We must ensure that words are matched with actions to evolve quality leadership and followership. A logical and strategic approach of immediate, short-term, medium-term, and long-term programs and projects implementation must be adopted.

In summary, the degree to which Nigeria will unleash it's technical and managerial capacity and capability to engage in innovation and specialization will depend on it taking the opportunities strategically based on results of a comprehensive system SWOT analysis in tandem with a functioning SETI system. This must mean that we are being innovative at product, process, managerial, and organizational levels. These are the levels that will guide the way forward for evolving a viable SD of the Nigerian economy.

What needs to happen is that Nigeria *must* nurture a favorable environment for the SETI to work. A strong industrialization strategy is needed with clear benchmarks and expected outcomes. Innovation support systems such as research funds, angel funds, industry venture capital, and technology risk funds must start to emerge. It is essential to make a commitment to create a critical mass of relevant skilled manpower to support strong management capabilities, such as feasibility studies, business planning, branding, increasing competitiveness, etc. In order for innovation to emerge, we must cultivate a culture of a strong IPR policy and management system. The education, training, and learning system must incorporate a strong technology entrepreneurship culture (incubation, Small and Medium Enterprises (SMEs), special competitive products and processes, high-level expertise).

One excellent review suggests that knowledge institutions such as universities must transform their operations to emerge not just as facilities for students to come in and graduate with certificates but such system must evolve

1. As a training center, to build communities and individuals' capacities to become self-reliant, entrepreneurs, and leaders of local development.
2. As a research, development, and experimentation center, to study and adapt local, national, and international technologies to local contexts and ensure their affordability for local communities and individuals.
3. As a production and industrial center, to test its development model and thus make its activities and discourses credible. It must develop and continuously refine an innovative system of production that is holistic, integrating primary, secondary, and services sectors;

competitive; and sustainable, building local human capacities and preserving natural resources.
4. As a structured enterprise, which needs to sell its products to survive.
5. As a service center, to support former trainees through a variety of services that continuously develop their capacities.
6. As a collaborator with national and international partners to widen its impact and also to work with the government on policies and programs concerning development, and to create new viable facilities in the countries.
7. To maintain and serve as a source of data, records, and archives for knowledge and development.

Furthermore, government activities should realize the requirement of heavy investments in the following key types of capitals:

1. Human capital, such as values, know-hows, managerial and organizational skills, to become innovative entrepreneurs through a deliberate but sustained program of reorientation and youth development.
2. Environmental capital, to recognize the value and potential of their natural environment—the sun, climate, rivers, seeds, insects, and so on—and organize them to produce, process, and commercialize effectively and efficiently, thus creating wealth.
3. Social/cultural/interpersonal capital, to network and create connections for sharing knowledge, experiences, difficulties, or opportunities, but also to negotiate, discuss, and work jointly.
4. Technical capital, to discover, develop, and incorporate innovative technologies—for example, new tools, machines, production techniques—to improve productivity.
5. Financial capital, to enable better management of their finances and start their own sustainable projects. Doing things differently in the sense that grants or loans and other financial supports proceed only if the four capitals above have been developed, to ensure that the beneficiaries are impregnated by the logic of productive production and innovation.

All these *capitals* have to be developed together to be effective, because they reinforce one another to maximize impact. These will encourage the natural overlap and connection between the SETI elements to evolve a functional SETI system for transformation. This is the environment that is commonly referred to as the triple helix where government, academia, and industry are all properly interlinked for SD.

When this overlap is functioning, clear performance indicators will emerge, including the following:

1. Affordable and dependable access to Information Communication Technology (ICT) facilities including the Internet for education, health, security, and social purposes
2. The libraries and laboratories in educational institutions transforming to be competitive and internationally standard learning and knowledge creation facilities
3. Physical infrastructure including roads; rail; electric power; and communication facilities mostly designed, developed, constructed, commissioned, operated, and managed by Nigerians because we have the knowledge and skills
4. Farming and agriculture transformed to be the hubs of development because we can produce the raw materials and add value at all levels to produce globally competitive consumable products
5. Evolving large pool of enterprises particularly at the SME level that are innovative with competitive branded products globally

Conclusions and recommendations

To conclude, it is evident from the discussions above that a strategy to transfer technology either by domesticating imported knowledge/know-how, or by moving the outputs of technological endeavors from the National Science, Engineering, Technology and Innovation System, is imperative for evolving a sustainable national development. A careful analysis of the outputs and potentials of the SETI system and initiation of programs whose projects can either be of immediate term, short term, medium term, or long term should immediately be launched in Nigeria. In line with this, the following are proposed as options of practical endeavors that can assist this to be achieved:

1. There must be a deliberate program to use existing technologies to grow viable middle class entrepreneurs in Nigeria. This would require knowledge workers and technology service providers to modify their operations and impacts through a revolutionary program that would fundamentally lead to the creation of a minimum 1,000,000 technical and managerial jobs using Nigerian engineering, technologies, and know-how (most of which are efforts having direct implications on the outputs of the SETI system). The strategy here should be on *building* the bridge needed to overcome the *valley of death* for technology.
2. To ensure that we build the required critical mass of highly skilled manpower, we must be very innovative and resolve to meaningfully

partner with technology service providers particularly those that transfer foreign technology to Nigeria. We must use the platform of implementing all capital projects in industry and at all levels of the public sector to build such technical capacities as MSc, MPhil, and PhD holders, and capabilities that would include quality engineering workshops, test stations, databases, and laboratories, to energize the knowledge system use quality platforms to acquire and/or domesticate technology. To achieve this, a creative platform of public and private partnerships must evolve.

3. We must mount deliberate programs to intensely and proudly recognize and promote technologies and innovations emerging from the National Innovation System, especially through viable partnerships, synergies, and networking of all related institutions on both supply and demand sides. Such promotion activities must be regular, qualitative, effective, and measurable.
4. We must institute a deliberate posture on *thinking out of the box* to acquire and transfer high-level technologies, to ensure that a clear linkage of academia to industry prevails. A typical tool on this is the establishment of knowledge-based functional science and technology parks infrastructure in all parts of the country.
5. Facilitate for the development of a patriotic and nationalistic posture for development through a deliberate program coined (in National Office of Technology Acquisition and Promotion [NOTAP]) as Agenda 177712. This agenda is based on specific projects that will demonstrate and deliver:
   a. A United Nigeria (the 1)
   b. That in the period of the next 7 years—2012 to 2019 (the 1st 7)
   c. That Nigeria will be identified with seven globally acclaimed quality branded products from its knowledge system (the 2nd 7)
   d. That seven multinational companies will emerge to ensure that the knowledge system in Nigeria will mature to a level strong enough to support industry and generate high-level IP for both local industrial consumption and for export (the 3rd 7)
6. The active contributions and roles of the Nigerian professionals in the diaspora are also critical. Against the well-known challenges of "Brain Drain," the energy of their technical expertise and participation in Nigeria is the "Brain Gain" for the country. Experiences from India, South Korea, China, and Malaysia to mention but a few point to the need for Nigeria to urgently improve its diaspora engagement initiatives by additionally adopting the following:
   a. Facilitate interaction between national professional associations and the diaspora professionals to encourage the flow and absorption of cutting-edge scientific research and technology

b. Establish avenues of collaboration between institutions of higher learning in Nigeria and the diaspora for mutual benefits
c. Facilitate the opportunities and encouragement for the diaspora to establish and own or sponsor institutions of skills/knowledge acquisition in critical areas of need, and also hi-tech companies
d. Online interaction opportunities to be established to engage with valuable diasporas who may not physically be able to come home
e. Source and facilitate exchange programs, scholarships, and internships both abroad and locally to maximize on the diaspora technology potentials

7. Evolve clear strategies that would engage the national professional institutions particularly The Academies of Science and Engineering to unleash the abundant technical energy available in them for speedy acquisition and application of relevant technologies to develop the Nigerian economy. This should include activities that would enable the full synergy and alignment of values between the academia, government, and industry for optimum and sustainable economic development.
8. For all the above to quickly happen, Nigeria must create the environment for a knowledge-based economy to emerge through the timely implementation of the recently approved Science Technology and Innovation Policy (Federal Ministry of Science and Technology 2012) particularly regarding the establishment and operation of the National Research and Innovation Fund aimed at pooling financial resources specially meant to support/boost the activities of scientific R&D in Nigeria. The fund should aim at achieving the minimum level equivalent to about 1% of the GDP as resolved by the African leaders at the African Union Summit of Heads of States in Addis Ababa in January 2007.
9. Also as approved in the new Science, Technology, and Innovation (STI) Policy, the National Research and Innovation Council chaired by His Excellency the President should urgently be functionalized, with similar structures at the state level chaired by the executive governors. These will facilitate for the needed synergy between all key stakeholders to position knowledge governance appropriately at the highest level, a feature commonly observed in most of the countries of the top 20 economies in the world. Implementing all the above in a systematic and logical manner will almost certainly ensure that Nigeria leverages on its SETI system, to evolve a sustainable economic development, well recognized and respected globally. This is a viable option for transforming the Nigerian economy to attain the Vision 20-2020, and it is also well aligned to the values and objectives of the Nigerian knowledge academies, especially the Nigerian Academy of Engineering. The academies therefore must act immediately!!!

References

Bindir, U. B., Lead Presentation, *The 5th Nigerian Universities Research and Development Fair*, Federal University of Technology Minna, Nigeria, October 9, 2012.

Federal Ministry of Science and Technology, *The National Science Technology and Innovation (STI) Policy*, Nigeria, 2012.

Hargroves, K., M. H. Smith, *The Natural Advantage of Nations: Business Opportunities, Innovation and Governance in the 21st Century*, Earthscan, London, 2005.

The National Planning Commission, *Nigeria Vision 20-2020, Economic Transformation Blueprint*, National Planning Commission, Nigeria, December 2009.

National University Commission (NUC), Workshop on the Application of the Nigerian University System Management Portal. *Monday Bulletin*, Vol. 8 No. 14, National University Commission, Abuja, Nigeria, April 15, 2013.

The Organisation for Economic Co-operation and Development (OECD), *21st Century Technologies: Promises and Perils of a Dynamic Future*, Chapter 2, 34pp. 1998.

chapter fifteen

Technology adoption*

Advanced technology has been dubbed the foundation of global industrial development. Proper transfer, implementation, and use of new technology are envisioned to produce a variety of products in quantities designed to satisfy local and regional demands in developing countries. Advances in electronic communication and business concepts as concurrent engineering and business process re-engineering, among others, could facilitate the rate at which products and processes that have been proven in industrialized nations, producing a win–win situation. This chapter aims to synthesize information on the technological needs of production systems for developing nations, using Nigeria as a case example. Using this needs assessment a framework will be developed, specifying the proper sequence and necessary procedures for technology adoption and the requirements for successful implementation of prescribed technologies. Utilization of existing human and natural resources in those countries as well as the necessary networks that must be developed with sources in the industrialized nations to make the process a success will be addressed.

Introduction

Experience has shown that in the ongoing push toward the global market place and ultimately a global village, there is a growing economic interdependence, thus making it less and less likely that any country will continue to survive and maintain a decent standard of living without exploiting and applying the appropriate technology to the fullest. Appropriate technology generally refers to machines or processes that will suit the countries they are intended to benefit. Valenti (1991) discusses specific case examples of engineers from highly industrialized nations taking their technology on the road to work with developing countries on the design of simple but effective machine that pumps water, cooks food, and transport the disabled. Komada (1986) argues that the issue of appropriateness

* Adapted with permission from Silvanus J. Udoka, and Emmanuel S. Eneyo, A framework for technology adoption in developing countries: Nigeria as a case example, in *Proceedings of the Second Africa-USA International Conference on Manufacturing Technology*, Nnaji and Badiru, editors, MANUTECH 1994, Sheraton Hotel Ikeja, Lagos, Nigeria, pp. 242–248, August 8–11, 1994.

of technology to a developing country is the most important issue in technology transfer (Geisel 1987). This issue has received great attention. This is because the technology is often capital-intensive and ill-suited to local production needs. The success or failure of technology transfer also depends on the ability of the receiving nation to identify the right technology for its needs. Todd and Simpson (1983) contend that the inappropriateness of certain technology transfers may be a result of dependence on existing technology for regulating development in developing countries. Technology that is structurally dependent has to be designed to suit the needs of receiving nation's experts and the experts from multinational corporation or nations, from where the technology is being imported this needs to arrive at an appropriate technology.

It is evident that economic growth and wealth have been created mainly in countries with considerable markets where large-scale production can be applied economically. Unfortunately, developing countries are increasingly at a disadvantage from an economic growth and industrial diversification standpoint, because of their small internal markets. There usually is not enough demand in most developing countries to justify the application of high-volume production technology (Brunak 1983). Manufacturers seeking to diversify and to develop new products usually find it difficult to economically justify local production due to the prohibitive cost of tooling required for a given product, relative to typically modest production volume. The process of identifying and selecting, and transferring appropriate technologies for adoption in developing countries can be effectively addressed only after adequate information has been accumulated on the state of technological development and needs of the target country, its political environment, and educational level of production workers.

Information requirements

One of the major causes of ineffective transfer of technology to developing countries is the lack of information and/or poor quality of available information on the available technology, relative success of their implementations as well as current and projected future technological needs. Developing countries have, in recent times, regularly spent considerable sums of money in development programs. Within these development programs, given governments' realization of the need for quantitative planning data for some or all sectors of the economy, a wealth of information is usually included for private firms, who are interested in performing a market analysis, a diversification study, or an examination of the capital facilities configuration (Beenhkker 1971). Consequently, the lack of data and poor quality information will increasingly be alleviated. In order for the transfer of the appropriate technology to developing countries to take place, continual improvements have to be made to ensure availability of quality information on a timely basis for technology could make the information available to manufacturing systems in developed nations readily available and applicable on a global scale. Computers, communication satellites, and an array of new and emerging information transmission technologies currently make teleconferencing and instant distant transactions a reality. The new information systems have the potential of linking manufacturing plants in developing countries with the sources of *know-how* in industrialized countries in an effective and affordable manner (Brunak 1983, p.81). The end result of this is an emerging concept of the *virtual manufacturing* aided by telecommunications and computers. Thus, by overcoming distance barriers, telecommunications, serving as a tool for the conveyance of information, is vital to technology development (Hudson 1987).

Distance would no longer be a disadvantage. Electronic communications would enable the required linkages for scores of people in disparate locations to work *side by side* through a globally integrated manufacturing and communication system. In such a setup, product selection would be achieved via teleconferencing; the implementation of production would be accomplished through automated retrieval manufacturing process plans from remote databases; training of production and maintenance personnel would be done through audiovisual tools in conjunction with other production equipment (Brunak 1983, p.82); and applications of such technologies as electronic data interchange and electronic graphics interchange in conjunction with data identifiers such as bar codes (Udoka 1993) could make a flow of manufacturing program data to computer-controlled machine tools and other equipment as well as a host of administrative functions to be achieved automatically and with minimum human interference and error.

Technological needs analysis

The foundation of any construct aimed at the transfer of technology into any developing country is the pool of information on the technological needs of the country. Using Nigeria as a case example, we would speculate that there are few, if any, industries with respect to which there exists today a comprehensive, unitary picture of technological needs of the production systems. The same is true of individual companies within any industry. Several individuals working for the same company may each have a perception and hence related information on the technological needs of their company. In all likelihood, no one within the company is assembling and systematically analyzing the information (Orbach 1983). There are, thus, very few to none of the manufacturing companies that can now plan for their technological development for the future years. This problem is exponentially compounded if considered at the national level. Technologies are continually being developed in the industrialized nations, and several of them now have the capability to affect several industries concurrently. These technologies need to be acquired, and personnel trained to use them parallel with the industrialized nations.

We propose that a complete needs analyses will be required for individual industries. Once the technological needs of individual industries are determined, a body of experts be set up to review all the single-industry analyses, compare the technology requirements to determine the commonality of the underlying technologies. Based on this overall analysis, the technological needs of every industry would be determined, pooling together, all of the information obtained into one comprehensive, easily accessible tool that will also identify the common needs, which could be met by the utilization of certain underlying technologies (Orbach 1983, p.45). This body would be expected to draw information from a database that had synthesized data from a number of sources, prominent of which would be: existing production systems that would be the recipients of the technology, research institutions that would investigate and study the process as well as the results of technology transfer, and from the government of the Federal Republic of Nigeria that should develop industrial development and technological transfer policy that will be conducive to, and encourage technological transfer. A synthesis of information from these and other pertinent sources would determine the technological needs of each industry and of the entire country.

Information repository on available technology

Once the national industry-focused technological needs have been defined, is it necessary to collect information about the available technology, and the transferability of the technology.

It is unlikely that any one company or industrial sector knows of the sources, from where to obtain relevant and comprehensive information to aid decision making when needed. For this reason, we again recommended a central repository of information as an essential component of technology transfer to Nigeria. This repository should serve as a data bank that is continuously upgraded with data from research and development bodies, individual industry sector sources, equipment manufacturers and distributors, and value-added resellers. We recommend that an Institute for Technological Advancement be set up in Nigeria and charged with the responsibility of maintaining an up-to-date information on available technology.

Matching the technology with the needs

It is logical to assume that for each technological need identified, there may be numerous technological systems responses. Of the various responses possible, it is also prudent to acknowledge the fact that some technological interventions will be more appropriate to the particular situation than others. The physical, economic, and social environment as well as the infrastructure for manufacturing differs significantly from one developing country to the next, and radically from those of the industrialized nations. For these reasons, not all technologies that are proven in industrialized nations are suitable as turn-key systems in the Nigerian production environment. A good example of this is the newspaper vending machine in the United States designed for the average honest customer. If, for instance, Nigeria is interested in this technology, this machine would have to be modified such that it dispenses one newspaper at a time, instead of allowing the customer full access to all the papers in the machine. This example clearly illustrates that the social structure of the importing country should be a prime factor in technology transfer and adoption. This demands the process be closely analyzed to selectively transfer only those technologies that are adaptable to Nigerian production environment economically and with minimal effort, and yet still maintain backward compatibility and connectivity with systems in the originating country to enable online communication should the need arise. Our recommendation is that the process should be set up to completely assess all potentially transferrable technologies that are capable of meeting specified technological need. Based on this, appropriate technology is selected that would best be integrated with existing systems, if any.

It should be recognized that all technological advances impact the environment. Along with the benefits of technological change are some undesirable side effects. Production technologies are a boon to socioeconomic environment because they provide for the production of high-quality products in an effective and efficient manner, while providing employment opportunities, among other benefits. These technologies also negatively impact the physical environment through air pollution, noise pollution, and in some cases, pollution of water supply. It is thus our recommendations that in the process of selecting the appropriate technology, significant strides have to be made to ensure that those selected for transfer and implementation in Nigeria will have minimal undesirable impact on the Nigerian environment.

Potential factors affecting technology transfer to Nigeria

Numerous factors individually or in various combinations restrict the rate at which technology can be transferred to developing countries, notably, Nigeria. We have already alluded to the lack of adequate information and/or relatively poor quality of available information. This is a key factor that must be properly addressed in order for there to be proper channels for transfer of technology to Nigeria. Other factors that would tend to hamper technology transfer are outlined in this section.

Another factor of significance is the lack of proper education and training of the personnel. It is commonplace for decision makers in developing countries to rise to management positions by promotions from shop-floor levels thus lack the proper level of formal education and are thus not properly equipped, and naturally will resist the adoption of complex technologies to address production problems. Our experiences of the Nigerian situation point to the fact that a shortage of well-trained personnel is not a serious problem as the misdeployment of well-educated individuals to responsibilities that are completely unrelated to their disciplines, and hence expertise. We suggest that the appropriate agencies seek out and match indigenes with relevant qualifications to the numerous opportunities for proper technology transfer and application in Nigeria. Indeed, this might spell an opportunity for the firms interested in transferring technologies in Nigeria to provide an incentive for enormous Nigerian expertise that is resident abroad to return either as consultant or members of the management team. Another approach to alleviate this problem would be to move in the direction of computerizing operations, thus requiring a few specialists, rather than a large number of semi-skilled and skilled workers (Beenhkker 1971, p.38).

Technical and administrative infrastructure

The lack of the proper infrastructure in developing countries poses another threat of expeditious transfer of technology from industrialized world. Most of the newer and advanced technologies operate in a computerized environment. Unfortunately, computers still remain status symbols rather than standard productivity-enhancing tools in developing countries. This implies that computerization and other pertinent technologies reside with the privileged few. Computerization and other technical and administrative infrastructural components should be considered as a vital part of the needs analysis. Given the rapid rate of changes occurring in the electronics industry, this phase of planning should avoid the impulse of planning in obsolescence.

Political instability

A significant number of companies and productive organizations in developing countries are government owned and controlled. There is also a tendency for significant influence on private and semiprivate organizations of governments in these countries (Beenhkker 1971, p.38). Because of the government's influence on the organizations, the research process often fails to reach the final stage of implementation. Generally, a project is dependent upon the organization's administration. Political instability, however, may result in replacing the organization's administration by the government, which interrupts the processes of development, one of which is the research process (Papageorgiou 1971, p.41). This thus creates a vicious cycle of the next administration realizing the need for advanced technologies, suggesting research and the processes outlined in this chapter, commissioning the project, starting the process, and investigating and implementating as interrupted, and the cycle repeats itself. These sequences of events result in the eventual delay in the development of the country and a whole gamut of negative attributes associated with it. It is our recommendation that while Nigeria as a nation is undergoing the learning curve and moving in the direction of economic and political stability, central organizations dealing with technology transfer and assistance to local companies should be set up to be instability proof. Indeed, they should have no political ties or allegiances and thus no political gains or losses in any political climate.

Win–win partnerships

New directions exist for economic growth that could produce a global economy that is far more responsive to basic human needs than those in existence. A variety of advanced technologies, a majority of which are nonresource-depletive, energy efficient, and overall environmentally friendly, have emerged in recent years and continue to be developed in industrialized nations. If judiciously transferred and applied in developing countries, they could bring about a major breakthrough in the fight to relieve severe poverty suffered by billions of the indigenes of developing nations around the globe. This could in turn produce substantial markets for equipment and associated auxiliaries, software, and international consulting opportunities for the industrialized nations, while creating jobs and helping improve the standards of living in developing nations.

Many small firms in the United States now face the reality that global trade is no longer a matter of choice, but rather one of survival. Consequently, growing numbers of small manufacturers are seeking foreign customers in their products. There is also continuing calls for increased co-production or production sharing of modern goods between industrialized and developing countries, instead of the advanced nations attempting just to sell (export) these goods to people who have barely enough to exist and thus cannot afford to buy them (Valenti 1991, p.83). There are thus remarkable advantages for both the source and recipient countries in the technology transfer endeavor.

The amount of preparations required for the production of a given product and mechanism for implementing the system to manufacture the product is enormous. The decision on the particular product to select for production at a given site will usually entail a search for a licensor, market analysis, technical evaluation, negotiations of contracts, gathering of information pertaining to manufacturing processes, equipment, manufacturing planning and control systems, materials requirements, and a host of other factors. These diverse tasks typically spread over a wide geographical area need to be accomplished in a timely and efficient manner, requiring modern information technology. All of these factors point to an enormous opportunity for international consulting opportunities and networks. These, again, would be advantageous to both the source and recipient countries.

Conclusion

With the attention of technological innovations and advancements as a means to increase productivity and maintain/improve the standard of living in industrialized nations, it is becoming more and more evident that these noble goals are no attainable if global economy continues to deteriorate, thus requiring international assistance and international policing and eventual urges to emigrate. It is thus imperative that strides be made to lessen the gap that separates industrialized from developing countries. One approach to accomplishing this is judicious transfer of proven technologies to developing countries from their industrialized counterparts. The process of identifying and selecting appropriate technology for transfer can be carried out only after significant information has been gathered, a technological needs analysis conducted, and information on technology that is available to satisfy the understanding of technology policy, industrial engineering and management, socioeconomic/socioenvironmental processes, and so on should be charged with the responsibility of matching the technological needs with sound technological answers, as well as providing the necessary networks/contacts between local industries and transfer sources. This approach holds the key to the quest to narrow the gap that separates developing countries such as Nigeria from the industrialized nations.

References

Beenhkker, H. I., Management sciences problems in developing countries, *Industrial Engineering*, pp. 37–39, 1971.

Brunak, B., CAM for developing nations, *Journal of Manufacturing Systems*, Vol. 2, No. 1, pp. 79–85, 1983.

Geisel, L. K., Technology transfer, knowledge engineering, and CIM, *CIM Review*, Vol. 3, No. 3, pp. 6–10, Spring 1987.

Hudson, H. E., Telecommunications and the developing world, *IEEE Communications Magazine*, Vol. 25, No. 10, pp. 28–33, 1987.

Komada, F., Japanese studies on technology transfer to developing countries: A survey, *The Developing Economies*, Vol. 24, No. 4, pp. 405–420, 1986.
Orbach, E., Technology transfer, *Productivity SA*, Vol. II, No. 4, pp. 44–46, 1983.
Papageorgiou, J. C., Why management sciences fail in developing countries, *Industrial Engineering*, pp. 40–42, 1971.
Todd, D., J. A. Simpson, The appropriate technology question in a regional context, *Growth and Change*, Vol. 14, No. 4, pp. 46–52, 1983.
Udoka, S. J., Electronic Data Interchange (EDI)/Electronic Graphics Interchange (EGI) and bar codes: Fundamental components of your world-class manufacturing enterprise, *Computers and Industrial Engineering*, Vol. 25, No. 1–4, pp. 139–142, 1993.
Valenti, M., Appropriate technology: Designing to fit local culture, *Mechanical Engineering*, pp. 64–69, 1991.

chapter sixteen

Technology transfer risk management

There is some level of risk in everything. This is particularly true in any technology-oriented endeavor. The dynamism and untested waters of new technology make risk very much a part of any technology undertakings. We should not shy away from risks. Rather, we should embrace risk as an opportunity to try new things so that we may thrive. Consider the following famous quotes:

> The future is always decided by those who put their imagination to work, who challenge the unknown, and who are not afraid to risk failure. (General Bernard A. Shriever, USAF, retired, 2005)

> It is not the critic who counts. Not the man who points out how the strong man stumbled or where the doer of deeds could have done better. The credit belongs to the man who is actually in the arena, whose face is marred by dust and sweat and blood; who strives valiantly; who errs and comes short again and again; who knows the great enthusiasms, the great devotions; who spends himself in a worthy cause. Who, at the best, knows in the end the triumph of high achievement, and who at the worst, at least fails while daring greatly, so that his place shall never be with those timid souls who know neither victory nor defeat. (Ted Roosevelt 1910)

The above quotes are about taking risk, venturing out, discovering what is out there, and exploring what exists within or outside the realm of possibility. Risk management is the process of identifying, analyzing, and recognizing the various risks and uncertainties that might affect a project. Change can be expected in any technology environment. It should be noted that change is the only constant in life. Change portends risk and uncertainty. Risk analysis outlines possible future events and their likelihood of occurrence. With the information from risk analysis, the technology transfer team can be better prepared for change with good planning and control actions. By identifying the various technology alternatives and their associated risk levels, the team can select the most appropriate courses of action.

Risk permeates every aspect of technology transfer and all metrics of technology management. Project scoping presents risks. Communication has risk components. Cost is subject to risk. Time has factors of risk and uncertainty. Quality variability contains a dimension of risk. Human resources pose operational risks. Procurement is subject to risk realities of the marketplace. Just as risk presents opportunities, it also poses threats. Thus, risk management is a crucial component of technology transfer and management. Risk is an uncertain event or condition that, if it occurs, has a positive or negative effect on at least one objective, such as time, cost, scope, or quality. In risk management, it is assumed that there are possible future states of a variable. Each occurrence of the variable has a known or assumed probability of occurring. There are often interdependencies in factors associated with a risk event. Thus, quantitative assessment is often very complex. Once a risk occurs, it is no longer a risk; it is a fact. There are three basic elements of risk given as follows:

1. There is some future event that has not occurred yet.
2. There is some level of uncertainty associated with the event.
3. There is a consequence (positive or negative) emanating from the risk event.

Risk management is the process of identifying, analyzing, and recognizing the various risks and uncertainties that might affect a project. The purpose of risk management is to achieve one of the following:

- Maximize the probability and consequence of positive events
- Minimize the probability and consequence of negative events

There are three possible risk response behaviors for risk management given as follows:

1. *Risk-averse behavior*: conscious and deliberate attempt to avoid risk
2. *Risk-seeking behavior*: conscious and deliberate pursuit of risk, perhaps as a manifestation of the old West saying that "you cannot accumulate if you don't speculate."
3. *Risk-neutral behavior*: indifference to the presence or absence of risk

For technology transfer purposes, it is imperative to understand and appreciate the role that risk management plays in the overall scheme of a manufacturing enterprise. Specific definitions, concepts, tools, and techniques of risk management are covered in this chapter.

Never try to avoid risk. Risk is the foundation for opportunity. The following list is a summary of what must be recognized in any technology transfer pursuit:

- Risk is always associated with a future event, where the time span between the present and the future may be as small as nanoseconds.
- Risk is an uncertain event or condition that, if it occurs, has an effect on at least one decision objective (e.g., cost, time, quality, performance, safety, and security).
- Risk may have one or more causes. They are requirement, assumption, constraint, and condition.
- Risk may have one or more impacts and/or outcomes.
- Risk impact and/or outcome may be negative or positive. A common misconception is that risk always implies a negative outcome. That is not the case.
  - Negative event implies a threat.
  - Positive event implies an opportunity.
- Risks exist as early as the beginning of a new endeavor.
- Risks may be known or unknown.
  - Known risks are those that have been identified and analyzed, making it possible to plan responses for those risks.
  - Unknown risks are those that cannot be managed proactively and should be addressed through a contingency plan.
- Risk that has occurred is considered an issue.
- Risk tolerance is the degree, amount, or volume of risk that an organization is willing to accept.
- Each technology transfer endeavor should develop a consistent approach to risk.
  - Communication about risk and its handling should be open and honest.
- The process of defining how to conduct risk management activities for technology transfer has the following characteristics:
  - Ensures that the degree, type, and visibility of risk management are appropriate for both the risks and importance of the technology to the organization
  - Provides sufficient resources and time for risk management activities
  - Establishes an agreed-upon basis for evaluating risks
    - Helps establish a plan of action for any risks that become concerns

Technology risk management plan

A technology risk management plan requires the following. The plan describes how risk management will be structured and performed.

- *Methodology*: Defines approaches, tools, and data sources that may be used to perform risk management
- *Roles and responsibilities*: Defines the lead, support, and risk management team members for each type of activity in the risk management plan, and clarifies responsibilities
- *Budgeting*: Assigns resources and estimates funds needed for risk management for inclusion in the cost performance baseline
- *Timing*: Defines when and how often risk management process will be performed
- *Risk categories*: Provides for a comprehensive process of systematically identifying risk to a consistent level of detail and contributes to the effectiveness and quality of the identified risks' process

SMART approach

The SMART (specific, measurable, aligned, realistic, and timed) approach to risk management involves the following practices:

- *Specific*: Risk management activities should be specific.
- *Measurable*: Risk management metrics should be measurable and actionable.
- *Aligned*: Risk management activities should be aligned achievable.
- *Realistic*: Risk management expectations should be realistic and relevant.
- *Timed*: Risk management actions should be timed. Time is of the essence in risk management. Action delayed is action misplaced.

Risk breakdown structure

Risk permeates everything in an organization. The overall risk management approach can be better managed by using a risk breakdown structure (RBS) similar to the conventional work breakdown structure (WBS) in project management. The specific elements of importance in the risk categories include the following:

- Technical risks
  - Technology attributes
  - Functional requirements
  - Quality of output
  - Performance assessment

- External risks
  - Suppliers
  - Contractors
  - Regulatory requirements
  - Industry standards
  - Business consensus
  - Customer requirements
  - Weather
  - Market climate
- Organizational risks
  - Hiring
  - Workforce development
  - Financing sources
  - Resource allocation
  - Management practices
- Technology management
  - Estimating
  - Planning
  - Controlling
  - Communication
  - Project phaseout strategies
  - Technology exit plan

Risk necessity

There are necessary and unnecessary risks. All endeavors involve a certain level of risk.

- *Necessary risk*: Acceptable risk means the potential benefits must outweigh the costs. We must accept only risks required to successfully complete a project. Even high-risk activities can be acceptable. Risk management is about controlling the level and nature of the risk, rather than avoiding all risks.
- *Unnecessary risk*: In this case, the potential return is not worth the potential loss. Unnecessary risk can occur when impatience, anxiety, pressure, temptation, and other impulsive lures are present in a decision environment. It is the responsibility of the decision maker to manage risk, plan ahead, set limits, and stay fully aware of the prevailing situation.

Levels of risk management

- *Deliberate risk management*: This is used for the formal pre-planning of all events or activities. Deliberate planning is used to focus on identifying anticipated hazards and developing effective strategies to mitigate the hazards. The planning phase can range from hours to months when resources are plentiful and time is available.

- *Real-time risk management*: This is an *on-the-fly* risk management approach that allows individuals to quickly recall, apply, and adapt risk management to changing conditions in *real-time*.

Risk handling options

There are several possible options for handling risk, including the ones listed below:

- *Reject the risk*: Reject a risk if the potential reward cannot outweigh the potential loss.
- *Avoid the risk*: It may be possible to avoid specific risks by going around the risk or doing the activity differently.
- *Delay the risk*: If there is no urgency to accomplishing a risky task, then a delay may be viable. Timing is of the essence in this case. Sometimes, issues resolve themselves over time.
- *Transfer the risk*: In this case, the risk is transferred to someone else, who may be better positioned to deal with the risk.
- *Spread the risk*: In this case, the risk can be shared with other risk takers or spread over a longer period of time to lessen a concentrated impact.
- *Compensate for the risk*: In this case contingency plans can be used to compensate for a risk. Creating a buffer, a backup, or a reserve is a good way to compensate for a risk.
- *Reduce the risk*: Risks can be reduced or avoided if hazards are identified in advance so that preemptive actions can be taken.

Risk probability and impact matrix

A probability and impact matrix provides the specific combinations of probability and impact that lead to a risk being rated as high, moderate, or low in importance. The corresponding levels of importance for planning responses to the risk are usually set by the organization. A probability and impact matrix has the following benefits:

- It prioritizes risks according to their potential impacts. One common approach uses a look-up table or a probability and impact matrix.
- It can revise stakeholder risk tolerances
- It defines how outcomes of the risk management processes will be documented, analyzed, and communicated.
- It documents how risk activities will be recorded for the benefit of the current effort as well as for future needs and lessons learned.

Risk factors

To be better technology risk managers, we need to understand risk factors. The primary factors to risk that always need to be considered on technology-related pursuits include the following:

- The risk event itself
- The probability of occurrence
- The range of possible outcomes

Other key considerations are timing and anticipation of the frequency of occurrence. Timing is an important consideration because the probability and impact change dramatically, depending on time of year and location of the effort. This is one area where global situational awareness is essential before embarking upon technology transfer arrangements.

- Risk management strategy
  - Includes assignments of roles and responsibilities.
  - Provides for risk management activities in the budget and schedule.
  - Identifies categories of risk.
- Cost estimates
  - Provides a quantitative assessment of the likely cost to complete scheduled activities.
- Activity duration estimates
  - Identifies risks related to time allowances for activities or the project.
- Project scope baseline
  - Provides project assumptions that should be evaluated as potential causes of risk.
- Stakeholder risk register
  - Identifies those impacted by the project and the concerns they may have.

Mitigating risk using the Triple C approach

Badiru (2008, 2009) present tools and techniques for managing projects in a complex environment, such as technology engagements. One technique is the Triple C approach to seeking and securing communication, cooperation, and coordination in project planning, execution, and control.

The Triple C model has been used effectively in practice to enhance project performance because most project problems can be traced to initial communication problems. The Triple C approach works because it is very simple, simple to understand and simple to implement. The simplicity comes from the fact that most of the required elements of the approach are already being done within every organization, albeit in a nonstructured manner. The Triple C model puts the existing processes into a structural approach to communication, cooperation, and coordination.

The qualitative approach of Triple C complemented the technical approaches used on the project to facilitate harmonious execution of tasks. Many projects fail when the stakeholders get too wrapped up into the technical requirements at the expense of qualitative requirements. The importance of Triple C is summarized in the sequential flow below:

Triple C → Communication → Cooperation → Coordination → Project Success

Other elements of "C," such as collaboration, commitment, and correlation, are embedded in the Triple C structure. The approach states that project management can be enhanced by implementing it within the integrated steps given below.

- Communication
- Cooperation
- Coordination

The model facilitates a systematic approach to project planning, organizing, scheduling, and control. The model places communication as the first and foremost function in any project. It addresses how the basic questions of what, who, why, how, where, and when revolve around the Triple C model. It highlights what must be done and when. It can also help to identify the resources (personnel, equipment, facilities, etc.) required for each effort right from the beginning. It points out important questions that are given as follows:

- Does each project participant know what the objective is?
- Does each participant know his or her role in achieving the objective?
- What obstacles may prevent a participant from playing his or her role effectively?

Triple C can mitigate disparity between idea and practice because it explicitly solicits information about the critical aspects of a project in terms of the following queries:

- Who
- What
- Why

- When
- Where
- How

Types of project communication

The most common types of project communication include the following:

- Verbal
- Written
- Body language
- Visual tools (e.g., graphical tools)
- Sensual (use of all five senses: sight, smell, touch, taste, hearing: olfactory, tactile, auditory)
- Simplex (unidirectional)
- Half-duplex (bidirectional with time lag)
- Full-duplex (real-time dialogue)
- One-on-one
- One-to-many
- Many-to-one

Types of cooperation

The most common types of project cooperation include the following:

- Proximity
- Functional
- Professional
- Social
- Power influence
- Authority influence
- Hierarchical
- Lateral
- Cooperation by intimidation
- Cooperation by enticement

Types of coordination

The most common types of project coordination include the following:

- Teaming
- Committee
- Delegation
- Supervision

- Partnership
- Token-passing
- Baton hand-off
- Working group
- Task force
- Technology action group

Typical Triple C questions

Questioning is the best approach to getting information for effective risk management. Everything should be questioned. By upfront questions, we can preempt and avert project problems later on. Typical questions to ask about technology transfer under the Triple C approach are as follows:

- What is the purpose of the technology?
- Who is in charge of the technology installation?
- Why is the technology needed?
- Where is the technology going to be located?
- When will the technology installation be carried out?
- How will the technology contribute to increased opportunities for the organization?
- What is the technology designed to achieve?
- How will the technology affect different groups of people within the organization?
- What will be the technology utilization protocol within the organization?
- What other groups or organizations will be involved (if any)?
- What will happen at the end of the life cycle of the technology?
- How will the technology impact be tracked, monitored, evaluated, and reported?
- What resources are required throughout the life cycle of the technology?
- What are the associated costs of the required resources?
- How do the technology implementation objectives fit the organizational goal?
- What respective contribution is expected from each participant?
- What level of cooperation is expected from each group?
- Where is the coordinating point for the new technology?

The key to getting everyone on board with technology transfer is to ensure that tasks and objectives are clear and comply with the principle of *SMART*.

Technology decisions under risk and uncertainty

Traditional decision theory classifies decisions under three different influences:

- *Decision under certainty*: This occurs when possible events or outcomes of a decision can be positively determined.
- *Decisions under risk*: This is made using information on the probability that a possible event or outcome will occur.
- *Decisions under uncertainty*: This happens when evaluating possible events or outcomes without information on the probability that the events or outcomes will occur.

Many authors make a distinction between decisions under risk and under uncertainty. In the literature, decisions made under uncertainty often incorporate decisions made under risk. Some of the parameters that normally change during a technology life cycle include cost, time requirements, technology performance specifications, and technology maintenance cost.

Cost uncertainties

In an inflationary economy, technology cost can become very dynamic and intractable. Cost estimates include various tangible and intangible components of a project, such as physical tools, inventory, training, raw materials, design, and personnel wages. Costs can change during a project for a number of reasons, including the following:

- External inflationary trends
- Internal cost adjustment procedures
- Modification of work process
- Design adjustments
- Changes in cost of raw materials
- Changes in labor costs
- Adjustment of WBS
- Cash flow limitations
- Effects of tax obligations

These cost changes and others combine to create uncertainties in the project's cost. Even when the cost of some of the parameters can be accurately estimated, the overall technology cost may still be uncertain due to the few parameters that cannot be accurately estimated.

Schedule uncertainties

Unexpected changes in requirements and other operational changes in the technology environment may necessitate schedule changes, which introduce uncertainties to the technology installation project. The following are some of the reasons project schedules change:

- Task adjustments
- Changes in scope of work
- Changes in delivery arrangements
- Changes in project specification
- Introduction of new technology

Performance uncertainties

Performance measurement involves observing the value of parameters during a project, and comparing the actual performance, based on the observed parameters, to the expected performance. Performance control then takes appropriate actions to minimize the deviations between actual performance and expected performance. Project plans are based on the expected performance of the project parameters. Performance uncertainties exist when expected performance cannot be defined in definite terms. As a result, project plans require a frequent review.

The technology management team must have a good understanding of the factors that can have a negative impact on the expected project performance. If at least some of the sources of deficient performance can be controlled, then the detrimental effects of uncertainties can be alleviated. The most common factors that can influence project performance include the following:

- Redefinition of project priorities
- Changes in management control
- Changes in resource availability
- Changes in work ethic
- Changes in organizational policies and procedures
- Changes in personnel productivity
- Changes in quality standards

Collection of risk management terms

Below is a collection of risk management terms, definitions, and explanations:

- *Risk*: Uncertain event or condition that, if it occurs, has a positive or negative effect on a project's objectives.

- *Risk acceptance*: Risk response planning technique that indicates that the project team has decided not to change the project management plan to deal with a risk, or is unable to identify any other suitable response strategy.
- *Risk avoidance*: Risk response planning technique for a threat that creates changes to the project management plan that are meant to either eliminate the risk or protect the project objects from its impact.
- *RBS*: Hierarchy organized depiction of the identified project risks arranged by risk category and subcategory that identifies the various areas and causes of potential risks; is often tailored to the specific project types.
- *Risk category*: Group of potential causes of risk; risk causes may be grouped into categories such as technical, external, organizational, environmental, or project management; category may include subcategorizes such as technical maturity, weather, or aggressive estimating.
- *Risk management plan*: Document describing how project risk management will be structured and performed on the project; it is contained in or is a subsidiary plan of the project management plan; information in the risk management plan varies by application and project size; is different from the risk register that contains the list of project risks, the results of risk analysis, and the risk responses.
- *Risk mitigation*: Risk response planning technique associated with threats that seeks to reduce the probability of occurrence or impact of a risk to below an acceptable threshold.
- *Risk register*: Document containing the results of the qualitative risk analysis, quantitative risk analysis, and risk response planning; risk register details all identified risks, including description, category, cause, probability of occurring, impact(s) on objectives, proposed responses, owners, and current status.
- *Risk tolerance*: Degree, amount, or volume of risk that an organization or individual will withstand.
- *Risk transference*: Risk response planning technique that shifts the impact of a threat to a third party, together with ownership of the response.
- *Sensitivity analysis*: Quantitative risk analysis and modeling technique used to help determine which risks have the most potential impact on the project; it examines the extent to which the uncertainty of each project element affects the objective being examined when all other uncertain elements are held at their baseline values; the typical display of results in the form of a tornado diagram.
- *Strengths, weaknesses, opportunities, and threats (SWOT) analysis*: Information-gathering technique that examines the project from the perspective of each project's SWOT to increase the breadth of the risks considered by risk management.

Conclusion

Risk management needs to be a part of every technology team's regular activities and should cover all key people, processes, business, and technology areas of the organization. A risk management plan helps the team to identify potential risks before they occur and prepares the team for a quick response if the risks occur. A well thought-out and proactive risk management plan can help the technology team to accomplish the following:

- *Reduce the likelihood that a risk factor will actually occur*: If only one person on the team fully understands the operation, losing that person in the middle of technology transfer could have serious adverse consequences. The team can reduce the risk by having a backup for each key expert and keeping documentation up-to-date and accessible.
- *Reduce the magnitude of loss if a risk occurs*: If the team suspects that the project has been underbudgeted, the team might be able to identify several backup sources to cover unexpected expenses.
- *Change the consequences of a risk*: A sudden reorganization, business acquisition, operational transition, or project termination in the middle of a project can seriously disrupt technology plans. If the team has established a process for dealing with abrupt changes, the team can meet the challenge with little or no impact to the overall technology goal.
- *Be prepared to mitigate risk during technology deployment*: The technology team can do this by strategically planning the technology project following the techniques presented throughout this book. Having technology backups and/or alternatives is the key to technology transfer success.

To minimize the effect of uncertainties in technology transfer, good control must be maintained over the various sources of uncertainty discussed earlier. The same analytic tools that are effective for one category of uncertainties should also work for other categories.

References

Badiru, A. B., *Triple C Model of Project Management: Communication, Cooperation, and Coordination*, CRC Press, Boca Raton, FL, 2008.

Badiru, A. B., *STEP Project Management: Guide for Science, Technology, and Engineering Projects*, CRC Press, Boca Raton, FL, 2009.

Roosevelt, Ted, "Citizenship in a Republic," http://en.wikipedia.org/wiki/Citizenship_in_a_Republic (accessed May 13, 2015), 1910.

Schriever, B., "Bernard Schriever Quotes," http://en.wikipedia.org/wiki/Bernard_Adolph_Schriever (accessed May 13, 2015), 2005.

Glossary

activity a component of work performed during the course of a project.

activity-based costing (ABC) bottom-up estimation and summation based on material and labor required for activities making up a project.

activity duration the time in calendar units between the start and finish of a schedule activity.

activity resource estimation the process of estimating the types and quantities of resources required to perform each schedule activity.

activity sequencing the process of identifying and documenting dependencies among schedule activities.

authority the right to apply project resources, expend funds, make decisions, or give approvals.

bar chart a graphic display of schedule-related information. In the typical bar chart, schedule activities or work breakdown structure components are listed down the left side of the chart, dates are shown across the top, and activity durations are shown as date-placed horizontal bars. Also called a Gantt chart.

baseline the approved time phased plan (for a project, a work breakdown structure component, a work package, or a schedule activity), plus or minus approved project scope, cost, schedule, and technical changes. Generally refers to the current baseline, but may refer to the original or some other baseline. Usually used with a modifier (e.g., cost baseline, schedule baseline, performance measurement baseline, technical baseline).

baseline start date the start date of a schedule activity in the approved schedule baseline.

best practices processes, procedures, and techniques that have consistently demonstrated achievement of expectations and that are documented for the purposes of sharing, repetition, replication, adaptation, and refinement.

change control identifying, documenting, approving or rejecting, and controlling changes to the project baselines.

close project the process of finalizing all activities across all of the project process groups to formally close the project or phase.

common cause a source of variation that is inherent in the system and predictable. On a control chart, it appears as part of the random process variation (i.e., variation from a process that would be considered normal or not unusual), and is indicated by a random pattern of points within the control limits. Also referred to as random cause. Contrast with special cause.

configuration management system a subsystem of the overall project management system. It is a collection of formal documented procedures used to apply technical and administrative direction and surveillance to identify and document the functional and physical characteristics of a product, result, service, or component; control any changes to such characteristics; record and report each change and its implementation status; and support the audit of the products, results, or components to verify conformance to requirements. It includes the documentation, tracking systems, and defined approval levels necessary for authorizing and controlling changes. In most application areas, the configuration management system includes the change control system.

constraint the state, quality, or sense of being restricted to a given course of action or inaction. An applicable restriction or limitation, either internal or external to the project, that will affect the performance of the project or a process. For example, a schedule constraint is any limitation or restraint placed on the project schedule that affects when a schedule activity can be scheduled and is usually in the form of fixed imposed dates. A cost constraint is any limitation or restraint placed on the project budget such as funds available over time. A project resource constraint is any limitation or restraint placed on resource usage, such as what resource skills or disciplines are available and the amount of a given resource available during a specified time frame.

contingency reserve the amount of funds, budget, or time needed above the estimate to reduce the risk of overruns of project objectives to a level acceptable to the organization.

control comparing actual performance with planned performance, analyzing variances, assessing trends to effect process improvements, evaluating possible alternatives, and recommending appropriate corrective action as needed.

control chart a graphic display of process data over time and against established control limits, and that has a centerline that assists in detecting a trend of plotted values toward either control limit.

control limits the area composed of three standard deviations on either side of the centerline, or mean, of a normal distribution of data plotted on a control chart that reflects the expected variation in the data.

cost control the process of influencing the factors that create variances, and controlling changes to the project budget.

cost of quality (COQ) determining the costs incurred to ensure quality. Prevention and appraisal costs (cost of conformance) include costs for quality planning, quality control (QC), and quality assurance to ensure compliance to requirements (i.e., training, QC systems). Failure costs (cost of nonconformance) include costs to rework products, components, or processes that are noncompliant, costs of warranty work and waste, and loss of reputation.

cost performance index (CPI) a measure of cost efficiency on a project. It is the ratio of earned value (EV) to actual costs (AC). CPI = EV/AC. A CPI value equal to or greater than one indicates a favorable condition and a value less than one indicates an unfavorable condition.

cost-plus-fee (CPF) a type of cost reimbursable contract where the buyer reimburses the seller for seller's allowable costs for performing the contract work and seller also receives a fee calculated as an agreed upon percentage of the costs. The fee varies with the actual cost.

cost-plus-fixed-fee (CPFF) contract a type of cost-reimbursable contract where the buyer reimburses the seller for the seller's allowable costs (allowable costs are defined by the contract) plus a fixed amount of profit (fee).

cost-plus-incentive-fee (CPIF) contract a type of cost-reimbursable contract where the buyer reimburses the seller for the seller's allowable costs (allowable costs are defined by the contract), and the seller earns its profit if it meets defined performance criteria.

cost-plus-percentage of cost (CPPC) see *cost-plus-fee.*

cost-reimbursable contract a type of contract involving payment (reimbursement) by the buyer to the seller for the seller's actual costs, plus a fee typically representing seller's profit. Costs are usually classified as direct or indirect costs. Direct costs are costs incurred for the exclusive benefit of the project, such as salaries of full-time project staff. Indirect costs, also called overhead or general and administrative cost, are costs allocated to the project by the performing organization as a cost of doing business, such as salaries of management indirectly involved in the project, and cost of electric utilities for the office. Indirect costs are usually calculated as a percentage of direct costs. Cost-reimbursable contracts often include incentive clauses where, if the seller meets or exceeds selected project objectives, such as schedule targets or total cost, then the seller receives from the buyer an incentive or bonus payment.

cost variance (CV) a measure of cost performance on a project. It is the algebraic difference between earned value (EV) and actual cost (AC). CV = EV – AC. A positive value indicates a favorable condition and a negative value indicates an unfavorable condition.

crashing a specific type of project schedule compression technique performed by taking action to decrease the total project schedule duration after analyzing a number of alternatives to determine how to get the maximum schedule duration compression for the least additional cost. Typical approaches for crashing a schedule include reducing schedule activity durations and increasing the assignment of resources on schedule activities. See also *fast tracking*.

create work breakdown structure (WBS) the process of subdividing the major project deliverables and project work into smaller, more manageable components.

critical activity any schedule activity on a critical path in a project schedule. Most commonly determined by using the critical path method. Although some activities are *critical*, in the dictionary sense, without being on the critical path, this meaning is seldom used in the project context.

critical chain method a schedule network analysis technique that modifies the project schedule to account for limited resources. The critical chain method mixes deterministic and probabilistic approaches to schedule network analysis.

critical path generally, but not always, the sequence of schedule activities that determines the duration of the project. Generally, it is the longest path through the project. However, a critical path can end, as an example, on a schedule milestone that is in the middle of the project schedule and that has a finish-no-later-than imposed date schedule constraint. See also *critical path method*.

critical path method (CPM) a schedule network analysis technique used to determine the amount of scheduling flexibility (the amount of float) on various logical network paths in the project schedule network, and to determine the minimum total project duration. Early start and finish dates are calculated by means of a forward pass using a specified start date. Late start and finish dates are calculated by means of a backward pass, starting from a specified completion date, which sometimes is the project early finish date determined during the forward pass calculation.

decision tree analysis the decision tree is a diagram that describes a decision under consideration and the implications of choosing one or another of the available alternatives. It is used when some future scenarios or outcomes of actions are uncertain. It incorporates probabilities and the costs or rewards of each logical path of events and future decisions, and uses expected monetary value analysis to help the organization identify the relative values of alternate actions. See also *expected monetary value (EMV) analysis*.

decomposition a planning technique that subdivides the project scope and project deliverables into smaller, more manageable components, until the project work associated with accomplishing the project scope and providing the deliverables is defined in sufficient detail to support executing, monitoring, and controlling the work.

defect an imperfection or deficiency in a project component where that component does not meet its requirements or specifications and needs to be either repaired or replaced.

defect repair formally documented identification of a defect in a project component with a recommendation to either repair the defect or completely replace the component.

deliverable any unique and verifiable product, result, or capability to perform a service that must be produced to complete a process, phase, or project. Often used more narrowly in reference to an external deliverable, which is a deliverable that is subject to approval by the project sponsor or customer. See also *result*.

Delphi technique an information-gathering technique used as a way to reach a consensus of experts on a subject. Experts on the subject participate in this technique anonymously. A facilitator uses a questionnaire to solicit ideas about the important project points related to the subject. The responses are summarized and are then recirculated to the experts for further comment. Consensus may be reached in a few rounds of this process. The Delphi technique helps reduce bias in the data and keeps any one person from having undue influence on the outcome.

develop project charter the process of developing the project charter that formally authorizes a project.

discrete effort work effort that is directly identifiable to the completion of specific work breakdown structure components and deliverables, and that can be directly planned and measured. Contrast with apportioned effort.

dummy activity a schedule activity of zero duration used to show a logical relationship in the arrow diagramming method. Dummy activities are used when logical relationships cannot be completely or correctly described with schedule activity arrows. Dummy activities are generally shown graphically as a dashed line headed by an arrow.

early finish date (EF) in the critical path method, the earliest possible point in time on which the uncompleted portions of a schedule activity (or the project) can finish, based on the schedule network, logic, the data date, and any schedule constraints. EVs can change as the project progresses and as changes are made to the project management plan.

early start date (ES) in the critical path method, the earliest possible point in time on which the uncompleted portions of a schedule activity (or the project) can start, based on the schedule network logic, the data date, and any schedule constraints. Early start dates can change as the project progresses and as changes are made to the project management plan.

earned value (EV) the value of completed work expressed in terms of the approved budget assigned to that work for a schedule activity or work breakdown structure component. Also referred to as the budgeted cost of work performed (BCWP).

earned value management (EVM) a management methodology for integrating scope, schedule, and resources, and for objectively measuring project performance and progress. Performance is measured by determining the budgeted cost of work performed (i.e., earned value} and comparing it to the actual cost of work performed (i.e., actual cost}. Progress is measured by comparing the earned value to the planned value.

earned value technique (EVT) a specific technique for measuring the performance of work for a work breakdown structure component, control account, or project. Also referred to as the earning rules and crediting method.

effort the number of labor units required to complete a schedule activity or work breakdown structure component. Usually expressed as staff hours, staff days, or staff weeks. Contrast with duration.

enterprise a company, business, firm, partnership, corporation, or governmental agency.

enterprise environmental factors any or all external environmental factors and internal organizational environmental factors that surround or influence the project's success. These factors are from any or all of the enterprises involved in the project, and include organizational culture and structure, infrastructure, existing resources, commercial databases, market conditions, and project management software.

execute directing, managing, performing, and accomplishing the project work; providing the deliverables; and providing work performance information.

expected monetary value (EMV) analysis a statistical technique that calculates the average outcome when the future includes scenarios that may or may not happen. A common use of this technique is within decision tree analysis. Modeling and simulation are recommended for cost and schedule risk analysis because it is more powerful and less subject to misapplication than expected monetary value analysis.

expert judgment judgment provided based upon expertise in an application area, knowledge area, discipline, industry, and so on. as appropriate for the activity being performed. Such expertise may be provided by any group or person with specialized education, knowledge, skill, experience, or training, and is available from many sources, including other units within the performing organization; consultants; stakeholders, including customers, professional and technical associations; and industry groups.

failure mode and effect analysis (FMEA) an analytical procedure, in which each potential failure mode in every component of a product is analyzed to determine its effect on the reliability of that component and, by itself or in combination with other possible failure modes, on the reliability of the product or system and on the required function of the component; or the examination of a product (at the system and/or lower levels) for all ways that a failure may occur. For each potential failure, an estimate is made of its effect on the total system and of its impact. In addition, a review is undertaken of the action planned to minimize the probability of failure and to minimize its effects.

fast tracking a specific project schedule compression technique that changes network logic to overlap phases that would normally be done in sequence, such as the design phase and construction phase, or to perform schedule activities in parallel. See also *schedule compression* and *crashing*.

finish-to-finish (FF) the logical relationship where completion of work of the successor activity cannot finish until the completion of work of the predecessor activity.

finish-to-start (FS) the logical relationship where initiation of work of the successor activity depends upon the completion of work of the predecessor activity.

firm-fixed-price (FFP) contract a type of fixed price contract where the buyer pays the seller a set amount (as defined by the contract), regardless of the seller's costs.

fixed-price-incentive-fee (FPIF) contract a type of contract where the buyer pays the seller a set amount (as defined by the contract), and the seller can earn an additional amount if the seller meets defined performance criteria.

fixed-price or lump-sum contract a type of contract involving a fixed total price for a well-defined product. Fixed-price contracts may also include incentives for meeting or exceeding selected project objectives, such as schedule targets. The simplest form of a fixed price contract is a purchase order.

float also called slack. See also *free float (FF).*

flowcharting the depiction in a diagram format of the inputs, process actions, and outputs of one or more processes within a system.

free float (FF) the amount of time that a schedule activity can be delayed without delaying the early start of any immediately following schedule activities.

Gantt chart see *bar chart.*

imposed date a fixed date imposed on a schedule activity or schedule milestone, usually in the form of a *start no earlier than* and *finish no later than* date.

influence diagram graphical representation of situations showing causal influences, time ordering of events, and other relationships among variables and outcomes.

integrated change control the process of reviewing all change requests, approving changes, and controlling changes to deliverables and organizational process assets.

invitation for bid (IFB) generally, this term is equivalent to request for proposal. However, in some application areas, it may have a narrower or more specific meaning.

lag a modification of a logical relationship that directs a delay in the successor activity. For example, in a finish-to-start dependency with a 10-day lag, the successor activity cannot start until 10 days after the predecessor activity has finished. See also *lead.*

late finish date (LF) in the critical path method, the latest possible point in time that a schedule activity may be completed based upon the schedule network logic, the project completion date, and any constraints assigned to the schedule activities without violating a schedule constraint or delaying the project completion date. The late finish dates are determined during the backward pass calculation of the project schedule network.

late start date (LS) in the critical path method, the latest possible point in time that a schedule activity may begin based upon the schedule network logic, the project completion date, and any constraints assigned to the schedule activities without violating a schedule constraint or delaying the project completion date. The late start dates are determined during the backward pass calculation of the project schedule network.

latest revised estimate see *estimate at completion.*

lead a modification of a logical relationship that allows an acceleration of the successor activity. For example, in a finish-to-start dependency with a 10-day lead, the successor activity can start 10 days before the predecessor activity has finished. A negative lead is equivalent to a positive lag. See also *lag*.

life cycle see *project life cycle*.

materiel the aggregate of things used by an organization in any undertaking, such as equipment, apparatus, tools, machinery, gear, material, and supplies.

matrix organization any organizational structure in which the project manager shares responsibility with the functional managers for assigning priorities and for directing the work of persons assigned to the project.

milestone a significant point or event in the project. See also *schedule milestone*.

Monte Carlo analysis a technique that computes, or iterates, the project cost or project schedule many times using input values selected at random from probability distributions of possible costs or durations, to calculate a distribution of possible total project cost or completion dates.

opportunity a condition or situation favorable to the project, a positive set of circumstances, a positive set of events, a risk that will have a positive impact on project objectives, or a possibility for positive changes. Contrast with threat.

organizational breakdown structure (OBS) a hierarchically organized depiction of the project organization arranged so as to relate the work packages to the performing organizational units. (Sometimes OBS is written as Organization Breakdown Structure with the same definition.)

parametric estimation an estimating *technique* that uses a statistical relationship between historical data and other variables (e.g., square footage in construction, lines of code in software development) to calculate an *estimate* for activity parameters, such as *scope*, *cost*, *budget*, and *duration*. This technique can produce higher levels of accuracy depending upon the sophistication and the underlying data built into the model. An example for the cost parameter is multiplying the planned quantity of work to be performed by the historical cost per unit to obtain the estimated cost.

Pareto chart a histogram, ordered by frequency of occurrence, that shows how many results were generated by each identified cause.

position description an explanation of a project team member's roles and responsibilities.

precedence relationship the term used in the precedence diagramming method for a logical relationship. In current usage, however, precedence relationship, logical relationship, and dependency are widely used interchangeably, regardless of the diagramming method used.

predecessor activity the schedule activity that determines when the logical successor activity can begin or end.

product life cycle a collection of generally sequential, nonoverlapping product phases whose name and number are determined by the manufacturing and control needs of the organization. The last product life cycle phase for a product is generally the product's deterioration and death. Generally, a project life cycle is contained within one or more product life cycles.

product scope the features and functions that characterize a product, service, or result.

product scope description the documented narrative description of the product scope.

program a group of related projects managed in a coordinated way to obtain benefits and control not available from managing them individually. Programs may include elements of related work outside of the scope of the discrete projects in the program.

program management the centralized coordinated management of a program to achieve the program's strategic objectives and benefits.

program management office (PMO) the centralized management of a particular program or programs such that corporate benefit is realized by the sharing of resources, methodologies, tools, and techniques, and related high-level project management focus.

project a temporary endeavor undertaken to create a unique product, service, or result.

project charter a document issued by the project initiator or sponsor that formally authorizes the existence of a project, and provides the project manager with the authority to apply organizational resources to project activities.

project life cycle a collection of generally sequential project phases whose name and number are determined by the control needs of the organization or organizations involved in the project. A life cycle can be documented with a methodology.

project organization chart a document that graphically depicts the project team members and their interrelationships for a specific project.

project scope statement the narrative description of the project scope, including major deliverables, project objectives, project assumptions, project constraints, and a statement of work, that provides a documented basis for making future project decisions and for confirming or developing a common understanding of project scope among the stakeholders. A statement of what needs to be accomplished.

resource leveling any form of schedule network analysis in which scheduling decisions (start and finish dates) are driven by resource constraints (e.g., limited resource availability or difficult-to-manage changes in resource availability levels).

responsibility matrix a structure that relates the project organizational breakdown structure to the work breakdown structure to help ensure that each component of the project's scope of work is assigned to a responsible person.

risk an uncertain event or condition that, if it occurs, has a positive or negative effect on a project's objectives. See also *risk breakdown structure (RBS).*

risk acceptance a risk response planning technique that indicates that the project team has decided not to change the project management plan to deal with a risk, or is unable to identify any other suitable response strategy.

risk avoidance a risk response planning technique for a threat that creates changes to the project management plan that are meant to either eliminate the risk or to protect the project objectives from its impact. Generally, risk avoidance involves relaxing the time, cost, scope, or quality objectives.

risk breakdown structure (RBS) a hierarchically organized depiction of the identified project risks arranged by risk category and subcategory that identifies the various areas and causes of potential risks. The risk breakdown structure is often tailored to specific project types.

rolling wave planning a form of progressive elaboration planning where the work to be accomplished in the near term is planned in detail at a low level of the work breakdown structure, while the work far in the future is planned at a relatively high level of the work breakdown structure, but the detailed planning of the work to be performed within another one or two periods in the near future is done as work is being completed during the current period.

root cause analysis an analytical technique used to determine the basic underlying reason that causes a variance or a defect or a risk. A root cause may underlie more than one variance or defect or risk.

schedule milestone a significant event in the project schedule, such as an event restraining future work or marking the completion of a major deliverable. A schedule milestone has zero duration. Sometimes called a milestone activity. See also *milestone.*

scope the sum of the products, services, and results to be provided as a project. See also *product scope statement.*

S-curve graphic display of cumulative costs, labor hours, percentage of work, or other quantities, plotted against time. The name derives from the S-like shape of the curve (flatter at the beginning and end, steeper in the middle) produced on a project that starts slowly, accelerates, and then tails off. Also a term for the cumulative likelihood distribution that is a result of a simulation, a tool of quantitative risk analysis.

statement of work (SOW) a narrative description of products, services, or results to be supplied.

strengths, weaknesses, opportunities, and threats (SWOT) analysis this information-gathering technique examines the project from the perspective of each project's strengths, weaknesses, opportunities, and threats to increase the breadth of the risks considered by risk management.

triple constraint a framework for evaluating competing demands. The triple constraint is often depicted as a triangle where one of the sides or one of the corners represent one of the parameters being managed by the project team.

value engineering (VE) a creative approach used to optimize project life cycle costs, save time, increase profits, improve quality, expand market share, solve problems, and/or use resources more effectively.

work breakdown structure (WBS) a deliverable-oriented hierarchical decomposition of the work, to be executed by the project team to accomplish the project objectives and create the required deliverables. It organizes and defines the total scope of the project. Each descending level represents an increasingly detailed definition of the project work. The WBS is decomposed into work packages. The deliverable orientation of the hierarchy includes both internal and external deliverables.

Index

Note: Locators followed by "*f*" and "*t*" denotes figures and tables in the text

F

U